Biophysics of the Senses
(Second Edition)

Biophysics of the Senses (Second Edition)

Tennille D Presley
Winston-Salem State University, Winston-Salem, NC, USA

IOP Publishing, Bristol, UK

ISBN 978-0-7503-3283-5 (ebook)
ISBN 978-0-7503-3281-1 (print)
ISBN 978-0-7503-3284-2 (myPrint)
ISBN 978-0-7503-3282-8 (mobi)

DOI 10.1088/978-0-7503-3283-5

Version: 20210401

IOP ebooks

British Library Cataloguing-in-Publication Data: A catalogue record for this book is available from the British Library.

Published by IOP Publishing, wholly owned by The Institute of Physics, London

IOP Publishing, Temple Circus, Temple Way, Bristol, BS1 6HG, UK

US Office: IOP Publishing, Inc., 190 North Independence Mall West, Suite 601, Philadelphia, PA 19106, USA

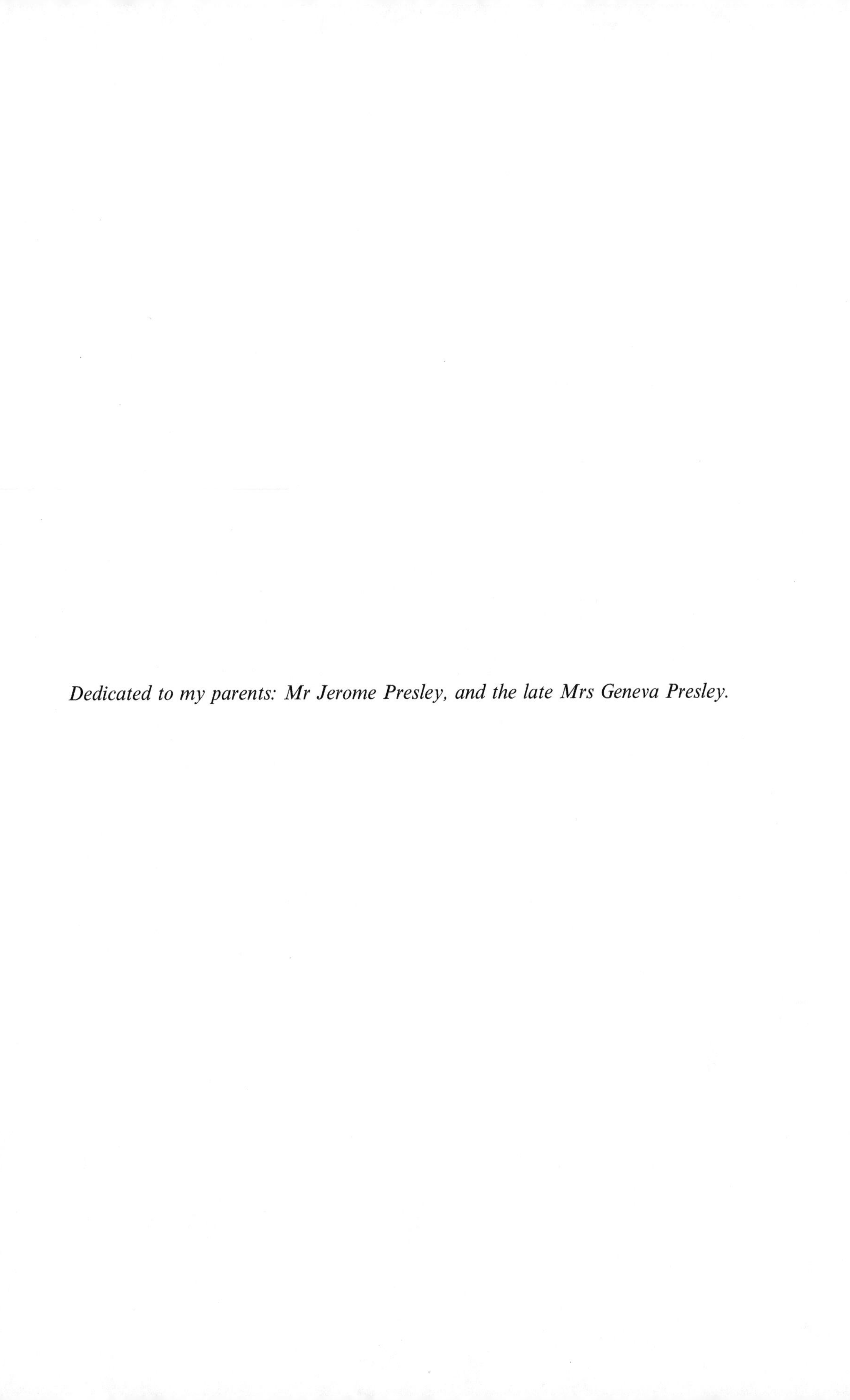

Dedicated to my parents: Mr Jerome Presley, and the late Mrs Geneva Presley.

Contents

Preface

I distinctly remember as a young girl being very fascinated by numbers and the desire to count money. At the age of two, I was able to count large sums of money and found great enjoyment in doing so. I was also delighted by the question of 'why'. Throughout my entire life, my dad would always fix things around the house. I enjoyed taking things apart and putting them back together—sometimes it would be right and other times I would have to ask my dad for help. As a high school senior, I took a physics class with an amazing teacher, who made the class very exciting and dynamic. Shortly after graduating from high school, my mother passed away and I found myself arriving back to the question of 'why'. It was this detrimental event that inspired my interest in studying the 'why' of the human body—'physics of the human body'. The question of whether there was a career that I could pursue to study the physics of the body arrived in my mind. At the time, I was unsure; however, I just knew that I enjoyed learning physics, was fascinated by how the body worked and was eager to learn more. Upon entering my interdisciplinary program as a freshman, I began to delve into the basic, fundamental principles of physics, biology and chemistry. My knowledge was further strengthened as I matriculated into my biophysics program. It was during graduate school that I knew that I was on the right path to decipher the aspects of 'why'.

I am often asked the question of 'What is the key or trick to physics?' My response is always: (1) pay attention to your units and thoroughly understand the units; (2) clearly understand your concepts to develop a strong foundation and uncover any vital, given information if you are trying to solve a word problem; (3) identify the question and determine the appropriate formula(s) to answer the unknown. Yes, the math is there and exists; however, you will not know what math to perform if you do not understand the concepts and the units. Remember that the units are just like your significant other, your siblings, parents, grandparents or children—you LOVE them! You should love your units the same way that you love the people that are near and dear to you. When you love your units, they will love you right back!

The tricks that I suggest to best tackle and understand physics involve the 'senses'. The most common senses associated with the human body are the senses of touch (tactioception), taste (gustaoception), sight (ophthalmoception), smell (olfacoception), and hearing (audioception). However, the human body also has the ability to respond to an array of stimuli such as variations in temperature, imbalance, excitement, pain and fear. Many would refer to these as a 'sixth sense'. On a daily basis, people migrate from place to place without stopping to think about the questions of *who, what, when, where and why—the five 'W's'. These are the essential 'senses' of biophysics that one must consider.* When applying these 'senses', one should always remember that the ideals and laws of physics can never be negated. They follow us everywhere that we go and aid in explaining how the nerve impulses send the signals within the body, leading to the known senses of sight, taste, touch, sound and smell. It is the five W's that provide clarity to the true existence of physics.

In this book, the reader will take a journey to look at the body from a physics perspective. Newton's laws of motion are applied to explain the mechanics of the body, whereas aspects of heat, energy and power elucidate how the body maintains a level of stability. The role of charges and free radicals in exercise and disease, and how music impacts the body from a biophysical perspective, are also addressed. This text is written for an undergraduate who may have an interest in medicine, one who may be new to physics, or one who may struggle with understanding why physics is important. Throughout the book, simple algebraic mathematics and the International System (SI) of units are used. Several questions of: 'how does physics impact a person's day-to-day life?', 'why is the body oriented the way that it is and why is it able to function in the manner that it does?' and 'why the body "vibes" a certain way to particular stimuli', will be addressed. Above all, this is my love story to physics and how understanding the field has provided a positive impact and improved my overall way of life.

Acknowledgements

First, I would like to thank God for allowing me the opportunity to create this piece of work.

I would like to thank Ms Kendra Royal for help with some of the figures, as well as Mr Jerome Presley for his insight on the '*Power Tools*' chapter.

I have a plethora of mentors and advocates that continuously inspire me and keep me motivated. I am eternally grateful to each of you, and value your wisdom and guidance. It is true that we as people stand on the shoulders of giants.

To my family and dearest friends, thank you for your love, support and prayers. You all help me stay grounded and focused. Particularly to my son, you keep me focused on my purpose.

Lastly to my students: Each of you that has ever taken a course with me, you challenge me to be creative and develop ways to make physics easy. Never stop learning.

Author biography

Tennille D Presley

Tennille D Presley, PhD is a tenured Associate Professor of Physics at Winston Salem State University (WSSU). She obtained her BS degree in Interdisciplinary Physics from North Carolina A&T State University, and acquired her MS and PhD degrees in Biophysics from The Ohio State University. While at The Ohio State University, she became the first African American to graduate from the Biophysics program, received the Young Investigator Award and the Best Advanced Research Award. Following her PhD, Dr Presley completed her post-doctoral training at Wake Forest University in the Department of Physics and the Translational Science Center. Since joining WSSU in 2010, she has been the recipient of numerous external and internal grants and awards including the Research Initiation Program for two years, the Preparing Critical Faculty for the Future Program Grant funded by the National Science Foundation and the Co-Director for the Provost's Scholars Science Immersion Program. Furthermore, she has been a part of the National Institutes of Health Programs to Increase Diversity Among Individuals Engaged (PRIDE) in Health Related Research division of Functional and Applied Genomics of Blood Disorders, Visiting Faculty at Brookhaven National Laboratory, a recipient of the Buckeyes Under 40 Award, and a US Delegate for the International Conference on Women in Physics. Most recently, her research has expanded into the realm of physics, music and biology as well as data science; she is a Faculty Fellow for the Center for Applied Data Science and NSF ASCEND. Dr Presley has an extensive record of national and international presentations, ultimately highlighting her scientific research and experience in biophysics, mentorship, advocacy and integrative learning. She is a well-sought out speaker and has published more than a dozen articles in free radical research.

IOP Publishing

Biophysics of the Senses (Second Edition)

Tennille D Presley

Chapter 1

Units: the essential tools to all understanding

Whether cooking, driving, or budgeting your expenses for the month, units play an important role in navigating a person's day-to-day life. They define the amount of something and are the driving force for providing guidance on the essential needs of an individual. Being the foundational basis for most things, it is imperative to become very knowledgeable about the assortment of units and how they apply to different scenarios. For instance, if a person travels from one country to the next, it is important to have the correct currency relevant to that particular country in order to properly function and perform necessary tasks. Ultimately, there is comfort is being educated on the universal language of units and diversifying this knowledge is key. In this chapter, the definition for units and their relevance is discussed; scientific notation, as well as the importance of unit conversion and the steps to convert units are also explored.

1.1 What are units and why are they important?

Universal Numeric Impartial Trusty Servants (UNITS)—these are the perfect words that embody the true essence and meaning for 'a unit'. Units are a measure of the amount of something and are designed to serve you. They are a universal language that can be well-trusted and are necessary to better understand the exact value of things encountered in a person's day-to-day life. A unit denotes a distinct magnitude for a standard measurement. For example, if you go to the bank and you request '100', chances are that you will be asked 'one hundred what'? Alternatively, someone may assume that you mean $100. However, this value could mean 100 pennies, 100 thousand dollars or even 100 shares of stock. It is evident that it is not only important to account for values, but to also specify a unit with the number to provide clarity. Units are the foundation to a true understanding, especially in the realm of physics.

When mentioning dollars or cents, a person automatically knows that those words refer to money. Irrespective of language or dialect, everyone speaks 'money'.

doi:10.1088/978-0-7503-3283-5ch1

The dollar is the American language for money and the amount can be converted to other types of currency. Similarly when discussing physical concepts, there are specific units that must be understood for maximum comprehension. Units are a person's saving grace when it comes to calculations in physics. There may be instances where a calculation or word problem may be confusing, but the units can be like 'tour guides' and lead you to the right place. They are the perfect navigation tools to solving a particular problem. If a person takes the time to learn their units and be consistent, then life in physics is far easier and more meaningful.

Units are also very important when cooking. If you are baking, it is imperative that the oven is set to the appropriate temperature in degrees Fahrenheit, that you monitor the amount of time for cooking (whether in hours or minutes) and that the ingredients are accurately measured to ensure the best taste. What if you are making a cake and the recipe states that you need a ½ dozen of eggs. Being aware that 1 dozen is equivalent to 12 individual eggs, you automatically know that ½ dozen accounts for 6 eggs. If ½ a cup of sugar is required, you automatically know that this represents 4 ounces of sugar to be measured. You can taste the foods and recognize if the adequate amount of ingredients were added. Having complete knowledge of what these values represent, promotes a more solid foundation of how the world works and functions. Similarly, it is known that a person has a pair of hands which equals 'two', three regions of the brain which represents a 'trio', and four chambers of the heart which is a 'quad'.

I always use the phrase 'If you love on your units, they will love you right back!' How much do you love your family and friends? Unconditionally, right? You should love your units the same way! Often a person may become frustrated with a problem and negate the units involved. However, units are very forgiving and will not waver from their job—their purpose is to 'serve' and guide a person in the right direction to the solution. When there is any doubt, it is imperative to know the specific units that directly correspond to individual concepts. In general, the standard system of units used by scientists is the International System (SI).

When considering the aspect of length, the meter is the *standard unit*. It was the first international standard ascertained back in the late 1700s, and was later redefined with respect to the origin of the speed of light. To put this into perspective, one meter is approximately the length from your shoulder to the tip of your index finger, whereas the length of a typical index finger is approximately 2½ inches. Commonly, the doorknob on a typical door is nearly one meter above the floor.

Each unit is unique in its own way, providing a clear indication of what concept is being discussed, as outlined in table 1.1. For instance, if someone says that they jogged 5 miles in 40 minutes, this reveals a lot of information, such as the distance that the person traveled (number of miles) and how long or the amount of time (number of minutes) that was necessary for the person to complete the five mile jog. The distance traveled per amount of time would be 0.125 miles/minutes. This defines speed, which we will discuss in a later chapter.

Measuring a value in particular units relies on making the distinction between both accuracy and precision. **Accuracy** is when a measurement is close to the exact value, whereas **precision** is the 'repeatability' of a measurement using the same

Table 1.1. List of common units.

Concept	Unit (s)
distance	miles (mi), feet (ft), meters (m), yards (yd), inches (in)
time	seconds (s), minutes (min), hours (h), days (dy), years (yr)
mass	grams (g)
weight	pounds (lb), Newtons (N)
volume	liters (L), cubic meters (m^3), cubic inches (in^3)
temperature	degrees Fahrenheit (°F), degrees Celsius (°C)

instrument or tool. Let's say that you are on your way to a birthday dinner; you put the address in the GPS (global positioning system) and the expected travel time is 15 min. As you drive, you arrive at the dinner within 12 min. Since you arrived 3 min ahead of schedule, the expected travel time was 'inaccurate'. Likewise, if you are moving into a new home or rearranging a room, it is necessary to measure the room and have the exact dimensions so that you can ensure all of your furniture will fit into the room as desired. One of my favorite sports to watch is basketball. As the coach, your goal is to have all of your players be both accurate and precise. When a player stands at the free throw line, their precision is demonstrated by their ability to shoot the ball the same exact way each and every time. The hope is that their technique will be accurate and that the ball will go into the hoop each time.

Prefixes for units are a common practice, and the primary prefixes within physics are the 'centi-', 'milli-', 'kilo-', 'micro-' and the 'nano-'. Any of these prefixes can be conjoined to a base concept. For example, 100 centimeters is equivalent to 1 meter. Similarly, 1000 grams equates to 1 kilogram. Each prefix provides a shorthand way to account for a particular value. For instance in table 1.2, there are a list of the metric prefixes with the values that each represents, as well as the abbreviation associated with each metric prefix. Numbers are often written in the 'power of ten' notation best described as **scientific notation**. Scientific notation is a convenient method for expressing numbers that are considered to be extremely small or large when written in standard decimal notation. So what exactly does it mean when a value is considered to be 'extremely large or small'? An example is if you have the number 0.000 000 0345 written in standard notation. This number would be best demonstrated in scientific notation as 3.45×10^{-8}. Now if you want to provide more clarity and specify the number as 3.45×10^{-8} m, it is obvious that this number is a representation of a unit of measure for length. The number could also be written using a prefix from table 1.2 as 34.5 nanometers.

What about the measurements affiliated with your body and bodily organs? Is it necessary to know the amount of space your organs occupy or even the mass they contain? There are nearly 78 organs within the human body, where some occupy more space than others. Table 1.3 includes some measurements that are associated with the body, where the heart and brain are amongst the larger organs in the body. However, the skin is the largest organ apart of the body, having a mass of ~11 000 grams (depending on the weight of the individual). Measurements are key

Table 1.2. List of basic prefixes.

Metric prefix	Abbreviation	Value
Hella	H	10^{27}
Yotta	Y	10^{24}
Zetta	Z	10^{21}
Exa	E	10^{18}
peta	P	10^{15}
tera	T	10^{12}
giga	G	10^{9}
mega	M	10^{6}
kilo	k	10^{3}
hecto	h	10^{2}
deka	da	10^{1}
deci	d	10^{-1}
centi	c	10^{-2}
milli	m	10^{-3}
micro	μ	10^{-6}
nano	n	10^{-9}
pico	p	10^{-12}
femto	f	10^{-15}
atto	a	10^{-18}
zepto	z	10^{-21}
yocto	y	10^{-24}

Table 1.3. List of common measurements.

Concept	Measurement
Average length of human hand	18.1 centimeters
Length of a football field	120 yards
Average human life expectancy	81 years
Mass of an electron	9.11×10^{-31} kilograms
Mass of a normal, human heart	300 grams
Weight of a normal, human brain	3 pounds
Total blood volume for 70 kg person	5.5 liters
Normal body temperature	98.6 degrees Fahrenheit
Average weight of skin in the body	10 900 grams

to aiding in normal maintenance and function of the body. For instance, you need to know what size clothing and shoes you wear in order to obtain the correct sizes. Alternatively, if you are ill, it is important to know what is a physiological temperature and the pharmacist must be knowledgeable in order to prescribe the appropriate dosage of medication.

1.2 Unit conversions

Have you ever been in the department store and paid cash for a purchase? If you give the cashier more money that the purchase costs, the cashier must then determine how much change you are owed. Let's say that you are owed $2.85; however, the cashier recognizes that they are out of dollars when they prepare to give the change back to you. As they look down at the coins in front of them, they remind themselves that four quarters or even ten dimes will equate to one dollar. This is an important lesson for ensuring that the conversion of units is well-defined. The cashier must have a clear understanding of the value of the coins in front of them and how each correlates to the dollar amount of change that is required to be returned to you. The secret to every unit conversion is to know what the conversion factor is for the concept that is being converted. A **conversion factor** is a ratio which portrays the correlation between two units. Many questions within everyday life can be answered by utilizing the conversion of units.

When setting up a unit conversion, there are a few basic steps that should be taken. First, identify the unit that needs to be converted. Next, determine the appropriate conversion factor to use for the preferred unit. Table 1.4 lists common conversion factors that are often useful. To begin to convert, it is easiest to set up the problem like a 'train track', as demonstrated in example 1.1. In the first block or the numerator, place the unit that needs to be converted. The next block should have the desired unit in the numerator and the unit you wish to discard in the bottom block

Table 1.4. List of some conventional conversion factors.

Distance/length
12 inches = 1 foot
1 mile = 1609 meters
1 mile = 5280 feet
1 meter = 39.37 inches
1 inch = 2.54 centimeters
Mass/Weight
1 kilogram = 9.8 Newtons
1 pound = 4.45 Newtons
1 kilogram = 2.21 pounds
Time
1 year = 365 days
1 day = 24 h
1 year = 3.156×10^7 s
1 h = 3600 s
Volume
1 liter = 1000 cubic centimeters
1 gallon = 4 quarts

or the denominator. This is necessary so that the unwanted units will 'cancel' out with each other, as any value divided by itself is equivalent to one. You can always confirm that the conversion is accurately set-up if the units properly cancel out.

Example 1.1. A woman decides to take a jog on a trail in the park that is 3 miles long. How many meters does she jog?

Solution: Knowing that 1 mile is equal to 1609 m (from table 1.4)[1],

$$(3\ \cancel{\text{miles}})\left(\frac{1609\ \text{meters}}{1\ \cancel{\text{mile}}}\right) = \mathbf{4,\ 827\ meters}$$

Example 1.2. A football player weighs 175 pounds, but desires to gain 8 more pounds of muscle over the course of the next few months. (a) How much weight would this be in Newtons? (b) What would be the player's body mass?

Solution: If the football player plans to gain 8 more pounds, the player's weight will be:

$$175\ \text{pounds} + 8\ \text{pounds} = 183\ \text{pounds}$$

(a) To convert this weight into Newtons,

$$(183\ \cancel{\text{pounds}})\left(\frac{4.45\ \text{Newtons}}{1\ \cancel{\text{pound}}}\right) = \mathbf{814.35\ Newtons}$$

(b) To determine the mass[2] of the player,

$$(814.35\ \cancel{\text{Newtons}})\left(\frac{1\ \text{kilogram}}{9.8\ \cancel{\text{Newtons}}}\right) = \mathbf{83.1\ kilograms}$$

Whether you are taking a road trip and trying to decide how long it will take to arrive at a certain destination or if you have a headache and are trying to figure out how many milligrams of aspirin to take, the measurement of units is critical to understand. The world is driven by units and the corresponding conversion factors. As we deepen our knowledge in future chapters, we will delve into concepts that have units which are derived from basic fundamental units. We will also utilize units to help solve word problems.

---

[1] The reason that the conversion factor of 1609 meters is in the numerator and 1 mile is in the denominator is so that the units of 'miles' can be properly cancelled out, thus leaving only the unit in 'meters'.

[2] Alternatively, the mass could be determined by converting the weight of 183 pounds to kilograms. Since there are 2.21 pounds equal to 1 kilogram, we would divide the 2.21 pounds into the 183 pounds to cancel the units. Thus, the answer would be 82.81 kilograms (~83 kilograms), which is very close to the value that was calculated above.

Further reading

Giancoli D C 2016 *Physics: Principles with Applications* (Boston, MA: Pearson)
Hewitt P G 2015 *Conceptual Physics* 12th edn (Glenview, IL: Pearson)

IOP Publishing

Biophysics of the Senses (Second Edition)

Tennille D Presley

Chapter 2

Mechanics of the body

Movement is an intricate process of life that has many different ways that it occurs. The manner in which things move can be very distinct based on individual properties. There are clear explanations that help us to understand 'how' objects move, as well as 'why' things move in the way that they do. Knowing that the Earth moves along with everyone on it, facilitates the thought that the body was designed to move; a lack thereof over an extended period of time can have detrimental effects. Thus, it is imperative to have a clear foundation of knowledge for the ways that movement is relevant. This chapter provides an introduction to movement and the basics of how and why movement occurs, from the key contributors of kinematics to the forces that influence processes. Furthermore, the quantification of movement is addressed and ways in which this information translates understanding of the human body and bodily movements. Each of these aspects contributes to enhancing our knowledge and comprehension of motion.

2.1 What is mechanics?

Without even thinking, people awake in the mornings and get out of the bed to go to work, school or wherever desired. Laying in a supine or horizontal position, individuals are able to fold and twist their bodies so that they are able to get out of the bed. Yes, this is a natural process; however, how is this possible? The human body is constantly impacted by movements whether it is from the Universe, galaxy, cells, or internal processes. All of this is due to the unique mechanics of the body. **Mechanics** is defined as movement by way of forces and energy. When one thinks of mechanics, two main branches come to mind: (1) *classical mechanics*; and (2) *quantum mechanics*. Classical mechanics focuses on macroscopic objects and how forces impact them, whereas quantum mechanics deals with the interaction of energy and matter of very small, nanoscopic materials. It is thought to supersede classical mechanics when it comes to subatomic or molecular levels. The relevance of quantum mechanics lies within the uncertainty principle, and the behaviors of

doi:10.1088/978-0-7503-3283-5ch2

electrons, protons, and other objects that are part of the atomic scale. In this chapter, our attention will be on *classical mechanics*, as it relates to the importance of mechanics to the human body. We will further discuss quantum mechanics in later chapters.

There are two main components of *classical* mechanics: (1) *dynamics*—this explains why things move the way that they do; (2) *kinematics*—this explains how things move. These concepts date back to Aristotle when he addressed both natural motion and violent motion. These aspects of motion define why objects tend to stay at rest, have a natural position that they strive to achieve, and why objects are impacted by an external force. I always tell students that 'You have a natural tendency to want to stay home and relax or do something fun; yet, you have a forced or imposed motion to come to class to further your knowledge and to fulfill the requirements of your major'. Objects move on the basis of a forced motion or a natural position or state. This is the basis for how the cells in our body move and interact with other cells. Think of how white blood cells rapidly move to protect against injury and infection, or even how insulin binds to receptors on the cell membrane to maintain the appropriate blood sugar levels. Cells move to keep us alive and well! Our bodies are regulated by the impact of mechanics.

2.2 Speed, velocity, and acceleration

Kinematics explain how things move on the basis of three main terms: (1) **speed** (s)—how fast something is moving; (2) **velocity** (v)—how fast and in what direction something is moving; and (3) **acceleration** (a)—the change in velocity with respect to the change in time. When I drive my car to work daily, I pay close attention to how fast the car is traveling so that I do not exceed the speed limit. Speed is a distinct measure of a distance (how far something travels) per time, where the common units are miles per hour (mph). However, when considering the body or physics-based problems, the primary units used are meters per second ($\mathrm{m\ s^{-1}}$). The distinction between the units is based on the fact that within the body, speeds occur on a smaller scale. As mentioned in chapter 1, there are approximately 1609 meters equal to 1 mile (1 mi = 1609 m). When runners compete in a marathon, it is imperative to pay close attention to their speed to monitor or predict how quickly they will complete the race. Speed is a concept that impacts our daily lives, often in instances where we are not always aware or carefully observe.

Commonly, the questions of 'how fast', 'how long' and 'how far' arrive in kinematics-based problems. 'How fast' correlates with the amount of speed or velocity. If you ask for 'how long', this indicates a query of the amount of time and inquiring 'how far' denotes the distance or the displacement. **Displacement** (Δx) is a measure of the location of an object from the original starting point, while **distance** (d) is the amount of ground that is covered or traveled. These questions play a vital role in guiding how to solve a problem and direct which formula(s) should be used to arrive at a particular answer. A formula is a way to show the relationship between various concepts, and often includes an 'equal sign'. For example, the formula for speed is as follows:

$$\text{speed} = \frac{\text{distance}}{\text{time}}$$

or

$$s = \frac{d}{t}$$

Notice that the variable '*s*' symbolizes speed, while '*d*' implies distance and '*t*' designates the time. With speed or any other formula, there should only be one 'unknown' variable missing in order to solve for the actual answer. Example 2.1 provides both the distance and time, thus the speed can be determined.

If we know how fast something is going, why is it important to also know the direction? Direction is an important factor in distinguishing between a vector versus a scalar. A **vector** is a quantity that has both a magnitude and direction, whereas a **scalar** denotes magnitude only. Speed is a scalar. If you were taking a road trip and needed to determine an estimated time of arrival, it is essential that not only the speed is taken into account but also the direction of travel is well noted. This is best defined as **velocity** which is a measure of displacement (*how far an object is from its starting point*) per time, or simply put, it is the speed and the direction. This vector quantity can be best identified by:

$$\text{velocity} = \frac{\text{displacement}}{\text{time}}$$

or

$$v = \frac{\Delta x}{\Delta t}$$

In the formula above, 'Δ' is the Greek letter *delta*, which means '*change in*'. When this symbol is present, it suggests that the initial value of a particular concept is subtracted by the final value of the concept (Δ = *final* – *initial*). In example 2.1, the velocity is quantified. The initial position is 20 m and the final position is 20 m as well. Thus, Δx = final position – initial position = 20 m – 20 m = 0 m. Irrespective of time, the velocity will be equivalent to 'zero' in this example because the displacement is zero. Velocity is also important for how the blood flows within the body related to systemic and pulmonary circulation. We will address blood flow in a later chapter.

Example 2.1. A student begins walking to her class from her dorm, which is nearly 30 meters away. After walking two-thirds of the way, she realizes that she left her notebook in her dorm room, so she turns around to get her book. If it takes her 5 min to make it back to her dorm, what is the speed of the student? What is her velocity?

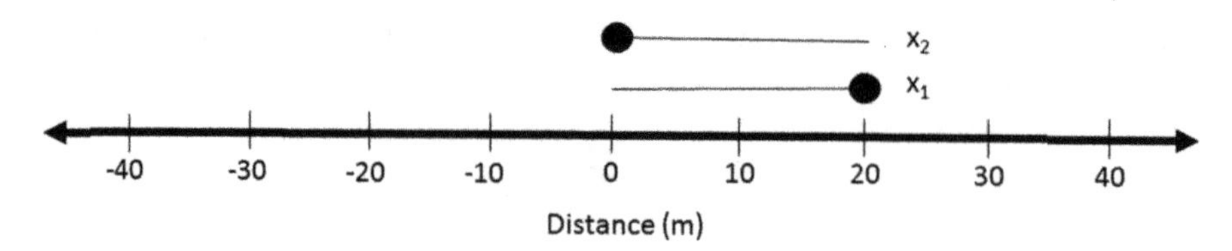

Figure 2.1. The student begins at the dorm (0 m) and returns back to her original position.

Solution: Let's first draw a picture that illustrates what is occurring in this problem. As shown in figure 2.1, the student begins at her dorm (at 0 meters) and walks ~20 meters, which will represent 'x_1'. She then turns around and walks back to the dorm, which represents 'x_2'. Thus, the total distance traveled is 40 m (20 m + 20 m = 40 m); however, the displacement is zero ($x_2 - x_1 = 20\text{ m} - 20\text{ m} = 0\text{ m}$), as she returns back to her original starting point. The time given is 5 min. Using figure 2.1 and the given information, we can determine both the speed and the velocity.

To determine the speed[1],

$$\text{speed} = \frac{\text{distance}}{\text{time}}$$

$$= \left(\frac{40\text{ m}}{5\text{ min}}\right)$$

$$\mathbf{Speed = 8\ m\ min^{-1}\ (or\ 0.133\ m\ s^{-1})}$$

To determine the velocity,

$$\text{velocity} = \frac{\text{displacement}}{\text{time}}$$

$$= \frac{0\text{ m}}{5\text{ min}}$$

$$\mathbf{velocity = 0\ m\ min^{-1}}.$$

Acceleration is the rate of change of an object's velocity and can be clearly elucidated with how an automobile works. If you are driving your car and you want to speed up the car, the gas pedal is pressed. This will result in an increase in the acceleration of the car. To slow the car down, you apply the brakes resulting in deceleration or a decrease in speed. To change the direction in which the car is moving requires turning the steering wheel. Based on this example, it reveals that an increase or decrease in speed, as well as altering the direction will ultimately impact an object's acceleration. Direction affects acceleration, thus making it a vector quantity, as well. If there is no change in velocity, then no acceleration occurs. The rate of change in velocity may be due to variations in speed and/or direction. Acceleration is represented as

$$\text{acceleration} = \frac{\text{velocity}}{\text{time}}$$

[1] To calculate the answer in 'm s^{-1}', simply convert using 1 min = 60 s.

Figure 2.2. Car accelerating.

or

$$a = \frac{\Delta v}{\Delta t}$$

The units for acceleration are meters per second squared (m s^{-2}). Velocity explains how quickly the position of an object changes, while acceleration tells how quickly the velocity changes. Example 2.2 asks us to determine *how fast* the car travels based on a given acceleration and an amount of time. Knowing that this question alludes to calculating the velocity, the formula for acceleration is used.

Example 2.2. A car accelerates (figure 2.2) down a street at a constant rate of 12.5 m s^{-2} within 2.8 s. How fast is the car traveling in miles per hour?

Solution: The given information is the acceleration of 12.5 m s^{-2} and the time equal to 2.8 s. Since the question asks 'how fast', this phrase implies that we should determine a speed or velocity. Hence, we will use the following equation:

$$\text{acceleration} = \frac{\text{velocity}}{\text{time}}$$

$$12.5\ \text{m s}^{-2} = \frac{\text{velocity}}{2.8\ \text{s}}$$

In order to determine the velocity[2], we must cross-multiply. Accordingly,

$$\text{velocity} = (12.5\ \text{m s}^{-2})(2.8\ \text{s})$$

$$\mathbf{velocity = 35\ m\ s^{-1}\ (or\ 78.3\ mph)}$$

Together, the three aspects of kinematics are the foundational formulas for understanding how movement occurs, especially in the body. Next we will address how the views of Sir Isaac Newton have impacted our way of thinking of mechanics.

2.3 Newton's laws of motion

Sir Isaac Newton is considered to be one of the most influential scientists of all time, especially when it comes to classical mechanics. Classical mechanics is often referred

[2] In order to convert the velocity from m s^{-1} to mph, we can first convert meters to miles by using the conversion factor of 1 mi = 1609 m; we know that there are 3600 s = 1 h. Thus, the tools gained for converting from chapter 1 can be used to determine that 35 m s^{-1} is equal to 78.3 mph.

to as 'Newtonian mechanics'. Newton revealed how every movement in the Universe can be mathematically analyzed. He is well-known for his three laws of motion. Newton's First Law of Motion is recognized as the *Law of Inertia.* **Inertia** is the tendency of an object to resist changes in motion and was first defined by Galileo; however, Newton refined this concept and made it his first law. The Law of Inertia simply states: 'An object at rest will stay at rest and an object in motion will stay in motion unless acted upon by an external force'. When referring to 'force', it is a push or pull, having the unit of a Newton (N). The Newton is equal to the product of a kilogram and meter per second squared (kg·m s^{-2}). Newton's first law is easily conveyed in a car with a seatbelt. If you approach a stop light and you have to decelerate the car by pushing the brakes, your body will lunge forward if you apply a great amount of force on the brakes. However, it will then return to its original, upright position. The body's desire was to resist the change in motion. Similarly, if you have a box at rest on the floor, the box will remain in that position unless 'something' or 'someone' (some type of force) causes the box to move.

Force is a vector and an important component in understanding physics and how things work. A stationary object needs a force to get it moving, while an object in motion needs a force to alter its velocity. There are a number of forces that are commonly influencing an object. Some include applied force, normal force, gravitational force, force of tension, air resistance, centripetal force, restoring force, buoyant force, and frictional force, all of which are described in table 2.1. Making a key distinction between the forces is essential to establish a clear understanding of a scenario, and some of these forces will be discussed in more detail in chapters to follow. Because force is a vector, the direction in which the

Table 2.1. List of forces.

Type of force	Description
Air resistance (F_{AR})	The opposing force relative to motion when an object travels through air
Applied (F_A)	The force that is applied or exerted by a person or another object
Buoyant (F_B)	The upward force that opposes the weight of an object when submerged in a fluid
Centripetal (F_C)	The force that acts on an object and towards the center when the object is moving in a circle
Friction (F_f)	The opposing force
Gravitational (F_g)	The weight of an object or the amount of force that Earth (or another massive object) attracts another object towards itself
Normal (F_N)	The upward or support force when an object is in contact with another object
Restoring (F_S)	The force that causes an object to return back to its equilibrium or original position
Tension (F_T)	The force exerted on a cable, cord, rope or string when pulled tight from opposing ends

forces act is important, and particularly when performing vector addition. Performing vector addition for two or more vectors is defined as the **resultant** (V_R). Objects that are parallel to each other or act along the same axis can be added to obtain the resultant of the vectors, knowing that vectors in the same direction will have the same sign (either '+' or '−') and those that are in different directions will have opposing signs. When perpendicular, vector addition is based on the Pythagorean Theorem ($a^2 + b^2 = c^2$):

$$V_R = V_1^2 + V_2^2$$

V_R accounts for the resultant of the vectors, while V_1 represents the first vector and V_2 is the second vector.

The best demonstration for showing the forces acting on a specific object is a free body diagram. A **free body diagram** is a visual representation of the forces acting on a particular entity, as shown in figure 2.3. In the figure, four forces are acting: normal force, applied force, frictional force, and the gravitational force. Friction is the resistive or opposing force that acts in the opposite direction of the applied force. There are two kinds of friction: static, which typically applies to things that are at rest, and kinetic, where the objects are in motion. Static friction is greater than kinetic. Think of it this way: if you have a heavy box that is sliding across the floor, it is usually easier to keep the box moving than if the box were originally stationary and then you attempt to move it. In general, an upward force exists, best known as the normal force or the support force. It is the force that counterbalances the gravitational force to provide stability. These forces are acting vertically in figure 2.3.

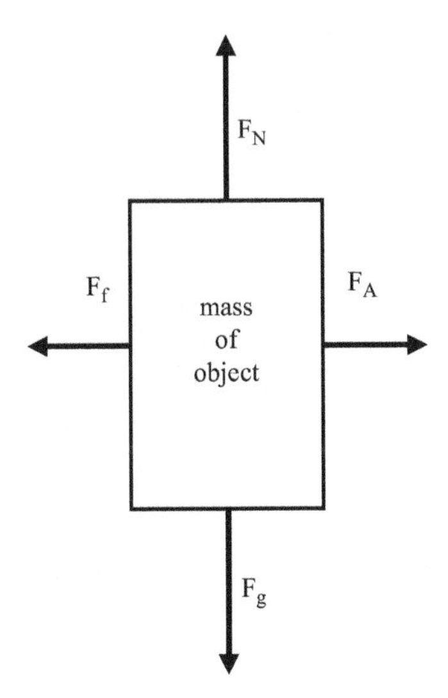

Figure 2.3. Example of a basic free body diagram, where the force of gravity, F_g, the force of friction, F_f, the applied force, F_A, and the normal force, F_N, act on an object containing a particular mass.

Example 2.3. Consider the free body diagram in figure 2.3. Assuming $F_f = 10$ N, $F_A = 4$ N, $F_N = 25$ N, and $F_g = 9$ N, determine the resultant for the following: (a) F_f and F_A; (b) F_N and F_g; (c) F_N and F_f.

Solution:

(a) F_f and F_A are acting along the same axes, but are in opposite directions. Thus, they are parallel vectors and their resultant would be obtained by:

$F_f = -10$ N (the 'negative' is because the force points towards the left)
$F_A = 4$ N
$V_R = -10\text{ N} + 4\text{ N} = \mathbf{-6\text{ N}}$

(b) Similar to part (a), F_N and F_g are acting in opposing directions along the same axis. Therefore, their resultant would be:

$F_N = 25$ N
$F_g = -9$ N (the 'negative' is because the force points down)
$V_R = 25\text{ N} + (-9\text{ N}) = \mathbf{16\text{ N}}$

(c) Unlike parts (a) and (b), F_N and F_f are perpendicular to each other. Considering, the following equation must be considered: $V_R = \sqrt{V_1^2 + V_2^2}$

$$
\begin{aligned}
F_N &= 25\text{ N}\\
F_f &= -10\text{ N}\\
V_R &= \sqrt{(25\text{ N})^2 + (-10\text{ N})^2}\\
&= \sqrt{625\text{ N}^2 + 100\text{ N}^2}\\
&= \sqrt{725\text{ N}^2}\\
&= 26.93\text{ N}
\end{aligned}
$$

Newton's second law is often acknowledged as the most popular of the three laws of motion and addresses the notion of force. Known as the *Law of Acceleration*, this law states that 'The sum of all forces or the net force is directly proportional to the acceleration and in the same direction as the net force acting on an object. The acceleration and the mass are inversely proportional to each other'. Newton's second law also means the net force of an object is equal to the rate of change of its linear momentum. Amongst the three laws, this one is represented by an equation:

$$\text{Net Force} = (\text{mass}) \times (\text{acceleration})$$

or

$$\text{Net Force} = \frac{\text{change in momentum}}{\text{change in time}}$$

In shorthand, it is often displayed as: $\Sigma F = ma$ or $\Sigma F = \frac{\Delta p}{\Delta t}$. In each equation, ΣF denotes the net force, m is the mass, a is the acceleration, Δp is the change in momentum and Δt is the change in time. 'Σ' is the Greek letter *sigma* and means the '*sum of*'. **Mass** (m) is a measure of an object's inertia or merely the amount of matter that makes up an object, having units of kilograms (kg). This concept does not change. However, a person or object's **weight** (w) is the product of the mass and the acceleration due to gravity (g) as shown below.

$$\text{weight} = (\text{mass}) \times (\text{acceleration due to gravity})$$

or

$$(w = mg)$$

Earlier in the chapter, the concept of acceleration was introduced. On Earth, the acceleration due to gravity (g) has a constant value of 9.8 m s^{-2}. This value can slightly vary depending on altitude. The gravitational pull on the Moon is nearly one sixth less than the Earth, resulting in an acceleration due to gravity of approximately 1.67 m s^{-2}; Jupiter has the highest acceleration due to gravity, which is roughly 25 m s^{-2}. Weight is a type of force (gravitational force) and has units of Newtons, like other forces that have been discussed. Especially in the United States, people measure their weight in pounds (lbs). A weight in pounds can be converted to Newtons by using one of the conversion factors that were mentioned in chapter 1. This is observed in example 2.4.

Example 2.4. An 11 pound poodle is lying on the porch (figure 2.4). As the mailman approaches the mailbox, the dog runs towards the mailman at 10 mph within 6 s. How much force does the poodle exert to get to the mailman?

Solution: The weight of the poodle is 11 pounds ($w = 11$ lbs), the initial velocity of the dog is 0 mph (since the dog is at rest), the final velocity of the dog is 10 mph ($\Delta v = 10$ mph) and the time is 6 s ($t = 6$ s). To determine the amount of force, Newton's 2nd law should be utilized.

Before we can use the formula, we must convert the velocity to m s^{-1}. The conversion factor for miles to meters is 1 mi = 1609 meters and 1 hour = 3600 s. Therefore, 10 mph = 4.47 m s^{-1} = Δv.

Next, the mass of the poodle should be quantified using the given weight. Let's use the formula for weight after converting the weight from pounds to Newtons. From chapter 1, we learned that 1 pound = 4.45 Newtons. Therefore, the weight is 48.95 Newtons.

$$w = mg$$
$$48.95 \text{ N} = m(9.8 \text{ m s}^{-2})$$
$$m = 4.99 \text{ kg}$$

Alternatively, the mass of the poodle could have been determined by using 1 kg = 2.21 lbs to obtain 4.98 kg (approximately 5.0 kg).

Figure 2.4. Poodle at rest, having an initial velocity of zero.

Based on what we are given, Newton's second law can now be applied to acquire the force:

$$\Sigma F = ma$$

$$\Sigma F = m\frac{\Delta v}{\Delta t}$$

$$\Sigma F = (4.99\ \text{kg})\frac{4.47\ \text{m s}^{-1}}{6.0\ \text{s}}$$

$$\mathbf{\Sigma F = 3.72\ N}$$

There are scenarios when the force of gravity is the only force acting. An instance where this occurs is when something is in *free* fall—the acceleration is equivalent to the acceleration due to gravity ($a = g$); thus, the acceleration is constant and the speed increases at 10 m s^{-1} for every second of fall. Objects may not only move left-to-right or up-and-down in a straight line, but may move diagonally or in a path that covers two dimensions. First described by Galileo, **projectile motion** is movement in two dimensions where an object travels in an arc near the Earth's surface only under the influence of gravity. The path of the horizontal component is at a constant velocity and the vertical component of motion is at constant acceleration. An illustration of projectile motion is when the quarterback passes the football on the field to the tight end or the receiver. As the ball is tossed into the air at a certain angle, it reaches a peak where the velocity along the vertical axis is zero. The ball continues to move with respect to being pulled downward by gravity.

As a lover of amusement parks, I am always intrigued by the amount of physics involved in the rides. **Circular motion** is a very common concept that exists. It occurs when an object is moving in a circle, and the speed is uniform but the direction of velocity changes. There are two main types of circular motion: centripetal (towards the center of the curve) and centrifugal (away from the center of the curve).

A centrifuge is an example of a centrifugal force. It is an instrumentation that is used to separate mixtures such as blood. It spins rapidly to separate blood into its three components—red blood cells, white blood cells/platelets and plasma. Centrifugal force is usually measured in 'g'. A typical rollercoaster exerts 2 g on the people who ride them. The ferris wheel and merry-go round are also popular rides that exhibit circular motion at the amusement park. When the ferris wheel rotates in a vertical circle, gravity as well as other forces that are acting on the body must be taken into consideration. How do you feel at the top of the ferris wheel compared to the bottom of the wheel? How do forces acting on the body at the top of the ferris wheel differ in comparison to how the forces act at the bottom of the wheel. Similarly, the merry-go round moves in a circle, but horizontally. Newton's second law of motion can address both of these scenarios, where centripetal force can be measured. It is understood that the gravitational and normal forces will act in the same direction towards the center of the circle when at the top of the ferris wheel ($\Sigma F = F_N + F_g$); however, at the bottom, the normal force will point towards the center of the circle and the gravitational force will point downward wheel ($\Sigma F = F_N - F_g$). Centripetal force is derived from Newton's second law and is represented as

$$\text{Net Force} = \frac{(\text{mass})(\text{velocity})^2}{\text{radius}}$$

This formula tells us that centripetal acceleration is equal to $\frac{\text{velocity}^2}{\text{radius}}$.

Newton's Third Law of Motion is the *Law of Action–Reaction.* It states 'For every action, there is an equal, opposite reaction'. An action–reaction pair acts on two different objects. As a child, did you ever play the game tug-of-war? You were demonstrating Newton's Third Law of Motion and probably did not know it. As you pulled the rope on your team, the opposing team was pulling the rope with an equal yet opposite amount of force. Together with the second law, the Law of Action–Reaction validates a number of the natural processes within the body, including how the body moves and is affected by motion. For instance, the second law addresses the change in **momentum**—the product of mass and velocity. The concept of momentum is often correlated with collisions. Let's say that you are walking in the grocery store and a child accidentally hits you with their shopping cart, causing a bruise on your leg. What amount of force does the shopping cart exert on you and you on it? The forces are equal, yet opposite. As the shopping cart pushes against your leg, your leg pushes back on the force. However, the cart's acceleration was greater.

2.4 Bodily movements

Thus far, we have discussed key aspects of mechanics and the basic forces that influence movement. How do these components affect the movement of the body? There are six basic bodily movements that people exhibit: walking/running, squatting, pushing/pulling, bending, lunging and twisting. Many of these movements are performed without even thinking about it. Our pair of hands, legs and feet allow us to fulfill most of these gestures. The musculoskeletal system is the primary system that impacts the mechanics of the body. There are three types of muscles in the human body: (1) cardiac; (2) skeletal; (3) smooth. Muscle cells are long, cylindrical cells that can contract when excited by nerve signals, allowing them to move bones. The contraction of muscles creates a force of tension on the bone and regulates bodily movement. Whether bending, squatting or lunging, muscle contraction occurs and force is present.

Skeletal muscles and joints are an intricate system of levers. A lever is a simple machine that lifts heavy objects. It consists of three main parts: (1) fulcrum—the pivotal point; (2) effort—the force applied to the end of the lever; (3) resistance—the opposing force on the other end of the lever. Within the body, the joints serve as the fulcrum, the force of the muscle is the effort and the body part's weight designates the resistance. Since muscles and joints are a system of levers and the applied force of the muscle onto the bone is rotational, the concept of torque is important. **Torque**—*the measure of how much a force acting on an object causes that object to rotate*—is required for an object to start or stop rotating. Torque must be understood when the body 'twists' and is represented by the following equation:

$$\text{Torque} = (\text{lever arm}) \times (\text{line of force})$$
$$\tau = r_{\perp} \cdot F$$

τ is the Greek letter 'tau' which denotes the torque in $\text{m} \cdot \text{N}$, $r_{\perp}$ is the perpendicular distance from the axis of rotation to the line of force with units of meters, and F is the applied force with the units of Newtons. Nodding your head, bending your arm or even standing on your tip toes are examples of torque. In each case, the forces acting are dependent on the position of the load from the pivot point.

Walking is a perfect example of when a person is out of balance for brief increments of time. As one foot is placed in front of the other, a force is applied to the ground and the ground exerts a force on us. In addition, the center of gravity extends beyond the foot that remains on the ground. **Center of gravity** is the point in the body where the average position of weight is properly distributed, and the *net* torque due to gravitational forces disappears. When an object is supported at its center of mass, it will remain in static equilibrium. Mechanical equilibrium is achieved when the net force is equal to zero. It represents a scenario where all of the forces are balanced—the amount of force in one direction is equivalent to the forces in the opposing direction. A perfect illustration is when a person is sitting in a chair. If the person does not lift vertically upward out of the chair or fall through the chair downward, then the net forces acting on the person are equal to zero and mechanical equilibrium is achieved. When a person is balanced, this means that the net force is zero and their body is in mechanical equilibrium. The ear is an important organ in the body which regulates equilibrium and balance. It is directly connected to the brain. Thus, the main function of the ear is not only to aid in hearing, but to provide equilibrium to the body, as well.

There are many factors that affect movement in our daily lives from various forces to how fast the movement occurs. The way that these aspects are involved is contingent upon the type of motion that occurs, whether it is one-dimensional or two-dimensional. It is important to maintain movement within the body and to also understand the impact of these movements (or lack thereof). So, let's keep moving!

Further reading

Fitts P M 1954 The information capacity of the human motor system in controlling the amplitude of movement *J. Exp. Psychol.* **47** 381

Giancoli D C 2016 *Physics: Principles with Applications* (Boston, MA: Pearson)

Herman I 2007 *Physics of the Human Body* (Berlin/Heidelberg: Springer)

Hewitt P 2015 *Conceptual Physics* 12th edn (Glenview, IL: Pearson)

McCall R P 2010 *Physics of the Human Body* (Baltimore, MD: Johns Hopkins University Press)

www.physicsclassroom.com

Chapter 3

How physics generates order in the body?

Managing and operating the body involves both external and internal processes. Whether you have experienced complete exhaustion, extreme joy or a common cold, a set of events occur in efforts to regulate order in the body. Understanding the meaning of the concepts involved and their contributions provides proper tutelage for having a more thorough grasp for how the body functions. Irrespective of the task, all bodily functions require energy. In this chapter, the way in which the body uses the five senses to maintain order is addressed. This aspect is linked to the intersection of work, power and energy, followed by the importance of thermodynamics, metabolism and pressure. Taken together, each of these concepts have a very distinctive way in which they contribute to generating and preserving order in the body; they work together in efforts to sustain health.

3.1 Bodily order

The body is a very complicated machine that thousands of individuals study to have a better understanding of how it works to prevent disease and to maintain a normal, functioning life. Despite many efforts, the body has its own innate processes for regulating a level of normalcy. We know that the body naturally lives and breathes when it is functioning properly. However, have you ever stopped and taken the time to ask how? There are numerous contributing factors that impact the normal day-to-day machinery of the body. **Homeostasis** is the regulation of internal conditions remaining constant and stable. The body's natural goal to maintain normal bodily function is due to homeostasis. This stability is impacted by the senses of sight, touch, taste, smell and hearing. While physics can be centered on each of these five senses; there is a 'sixth sense' that interposes the conventional operations of the body as well. This may be the voice in your head that tells you when something is wrong or that feeling that is in the pit of your stomach when you are on the right track to solving a problem. Either way, this helps to regulate your natural processes. This is the embodiment of physics!

doi:10.1088/978-0-7503-3283-5ch3

3.2 Work, energy, and power

As demonstrated in chapter 2, force plays an important role in the mechanics of the body. In the presence of force and displacement, work exists. If there is no force, then there is no work. Likewise, zero work transpires in the absence of displacement. Thus, work subsists only if there is motion involved. The units for work are the Joule, which is a Newton-meter (N·m). Work is defined as

$$\text{Work} = (\text{Force}) \times (\text{displacement}) \cos\theta$$

or

$$W = F\Delta \times \cos\theta$$

Work may be positive or negative, depending on the direction of the force and motion of the object involved. When the angle between the displacement and the force is less than 90°, then the work is positive. Positive work implies that both the direction of movement and force are the same. Alternatively, negative work is when the direction of movement and force act in opposite directions. An angle greater than 90° also leads to negative work. When the force and the displacement are perpendicular to each other ($\theta = 90°$), work is equal to zero. If you lift a gallon of milk from the shelf in the refrigerator, you perform work on the milk. However, if you hold the milk in a stationary position, there is no work being done because there is no displacement.

The ability to do work is defined as **energy**. All energy production originates in the cytosol of the cell. The five main forms of energy are chemical, mechanical, nuclear, thermal and electromagnetic. Each of these types have their own distinctiveness in regards to how they aid to regulate or contribute to bodily function. One of the key sources of chemical energy is from food. Mechanical energy is a composite of both **potential energy** (stored energy) and **kinetic energy** (energy due to motion). Potential energy comes in various forms including gravitational and elastic. Gravitational potential energy is with respect to the vertical position of an object, where the acceleration due to gravity must be considered. Elastic potential energy involves situations where energy is stored due to stretch. Think of our muscles and joints that we discussed in the previous chapter. As muscles stretch and move, elastic potential energy occurs. This 'spring' motion stretches the muscles and stores energy when force is applied, and recoils to release energy when force decays. Each of these types can be mathematically quantified using the formulas below:

$$\text{Potential Energy (gravitational)} = \text{mgh}$$

$$\text{Potential Energy(elastic)} = \frac{1}{2}kx^2$$

Gravitational potential energy (PE_G) is the product of mass (m), the acceleration due to gravity (g), and the vertical displacement or height (h). Elastic potential energy is directly proportional to the square of the amount of stretch, where k is the spring constant and x is the displacement. The spring constant represents the threshold of stretch. For a woman who has naturally curly hair, the curl in her hair is a perfect example of elastic potential energy. If the hair is stretched, the hair will tend to go back to its original compressed state once it is released. On the other hand, kinetic energy involves motion. The faster an object moves, the greater its kinetic energy will be. Considering, kinetic energy is indicated as

$$\text{Kinetic Energy} = \frac{1}{2}(\text{mass})(\text{velocity})^2$$

or

$$\text{KE} = \frac{1}{2}mv^2$$

The units of mass are the kilogram (kg) and the units for velocity are in meters per second (m s^{-1}). In a situation where the velocity doubles, the kinetic energy will enhance by a factor of 4.

Energy has the unique ability to be transformed amongst its different forms. This process occurs due to the **conservation of energy**—*energy is neither created nor destroyed, but simply transformed to another form; yet, the total amount of energy remains the same*. Efficiency is the capability of an object to complete a task based on the amount of energy supplied. Ideally, 100% efficiency is desired; yet, the efficiency of a person is nearly 25%. While work and energy have the same units, they are not the same thing. Work is a way to transfer energy. The work-energy principle is where the net work is equal to the change in energy ($W_{\text{net}} = \Delta\text{KE}$). This principle is a practical reformulation of Newton's laws of motion. When the net work is negative, the kinetic energy decreases; if the net work is positive, the kinetic energy is augmented.

Adenosine triphosphate (ATP) also provides energy. ATP is viewed as the molecular unit of 'currency' for energy transfer within the cell. Produced in the mitochondria, ATP is an unstable molecule, making it a 'universal' carrier of energy for the cells. For example, ATP is an energy source for muscle contraction, motility and active transport across the plasma membrane. ATP production is insufficient to sustain human life without the presence of oxygen. One meaningful role of oxygen is with the lungs. The main function of the lungs is to transport oxygen from the atmosphere into the bloodstream and to release carbon dioxide from the bloodstream into the atmosphere.

One afternoon, you decide to take a walk in the park. Before you know it, you have walked nearly three miles and did not feel tired; however, you may feel differently if you were to climb up three miles of stairs, as you are using more power. **Power** is the rate at which work is performed or the ratio of energy transformation with respect to time. This concept can be mathematically determined by using the following equation:

$$\text{Power} = \frac{\text{Work}}{\text{time}}$$

The units for power are the Watt (W) which is equivalent to a Joule per second (J s^{-1}). When a person thinks of power, it is often utilized in terms of its relevance to the electricity in a home or building. However, this same exact concept is pertinent to the normal operations of the body. If you were to run up a hill, you can measure your power based on your body's weight (the force), the height of the hill (the displacement) and the amount of time that it takes you to walk up the stairs. When walking versus running up the hill, the gravitational potential energy will remain the same; however, the power changes.

Example 3.1. While running a half marathon (13.1 miles), Latoya (w = 135 lbs) begins to feel tired. Halfway through the run, she expends 165 J within 1 h. (a) What is her mass? (b) How fast was she traveling? (c) How much power does she exert?

Solution:

(a) Since we are given the weight, we must convert from weight to mass. Referring back to chapter 1, 1 kilogram = 2.21 pounds. Thus,

$$(135\,\cancel{\text{lbs}})\left(\frac{1\ \text{kilogram}}{2.21\,\cancel{\text{lbs}}}\right)$$
$$= \mathbf{61.1\ kg}$$

(b) How fast indicates that the velocity must be determined. The kinetic energy of 165 J is given, and the mass is determined in part (a). Ultimately, $KE = \frac{1}{2}mv^2$ can be used to solve for 'v'. To solve,

$$KE = \tfrac{1}{2}mv^2$$

Plugging each value

$$165\ \text{J} = \tfrac{1}{2}(61.1\ \text{kg})v^2$$
$$165\ \text{J} = (30.55\ \text{kg})v^2$$

Solving for 'v'

$$\mathbf{v = 2.32\ m\ s^{-1}}$$

(c) To determine the power, we must use

$$\text{Power} = \frac{\text{Work}}{\text{time}}$$

Because the time is 1 h, it must be converted to 'seconds' which is 3600 s. In knowing that $W = F\Delta x$, the weight is equivalent to force and needs to be converted from pounds to Newtons. Using the conversion 1 pound = 4.45 Newtons, 135 lbs equals 600.75 N. Since Latoya travels halfway, the displacement is half 13.1 miles which equals 6.55 miles or 10 539 meters. Now plug in all of the necessary values

$$\text{Power} = \frac{F\Delta x}{\text{time}}$$

$$\text{Power} = \frac{(600.75\ \text{N})(10\ 539\ \text{m})}{3600\ \text{s}}$$

$$\boldsymbol{P} = 1758.\ \mathbf{7\ W}$$

3.3 Thermodynamics

Thermodynamics is the study of heat and its effects on temperature, pressure, volume, work, internal energy and entropy. Have you ever touched something and told yourself 'Wow that is hot or cold?' If the answer is 'yes', you have questioned the temperature of a substance. **Temperature** is simply a measure of how warm or cold something is. It is also noted as the amount of random kinetic energy of the molecules of a substance. Temperature may be ascertained with a thermometer in degrees Fahrenheit (°F), degrees Celsius (°C) or Kelvin (K). Both degrees Fahrenheit and Celsius directly measure temperature, but Kelvin measures energy. In chapter 1, we discussed unit conversions. Thus, this concept can be applied to these temperature scales since they can be converted amongst each other; this is highlighted in table 3.1. It is important to note that while degrees Fahrenheit and Celsius can be determined interchangeably by way of each of their equations, a given temperature in degrees Fahrenheit must be converted to degrees Celsius before it can be determined with respect to Kelvin. Example 3.2 shows how temperature can be quantified by using equations based on each temperature scale.

Example 3.2. The coronavirus (COVID-19) pandemic began disrupting the world in 2019, and affects individuals in very different ways. One of the key symptoms of COVID-19 is fever. As defined by the Centers for Disease Control and Prevention (CDC), a temperature of 100.4 °F is cause for concern. What is this temperature in both degrees Celsius and Kelvin?

Solution: Since we are given the temperature in degrees Fahrenheit (°F = 100.4), we can utilize the equations for both degrees Celsius and Kelvin.

Solving for °C,

Table 3.1. Temperature scales.

Temperature scale	Equation
Fahrenheit (°F)	$°F = \frac{9}{5}(°C + 32°)$
Celsius (°C)	$°C = \frac{5}{9}(°F - 32°)$
Kelvin (K)	$K = °C + 273$

$$
\begin{aligned}
{}^{\circ}\mathrm{C} &= \frac{5}{9}({}^{\circ}\mathrm{F} - 32^{\circ}) \\
&= \frac{5}{9}(100.4^{\circ} - 32^{\circ}) \\
&= \frac{5}{9}(68.4^{\circ}) \\
&= \mathbf{38}\ {}^{\circ}\mathrm{C}
\end{aligned}
$$

Next, we must determine the temperature Kelvin. (Note: it was necessary to determine the temperature in degrees Celsius first, since there is no equation to go directly from degrees Fahrenheit to Kelvin.)

Solving for K,

$$
\begin{aligned}
\mathrm{K} &= {}^{\circ}\mathrm{C} + 273 \\
&= (38) + 273 \\
&= \mathbf{311\ K}
\end{aligned}
$$

Heat (*Q*) is simply energy in transit from a higher temperature to a lower temperature. It can be quantified by accounting for the product of the mass (m), specific heat (c) and change in temperature (ΔT) of a material.

$$\text{Heat} = (\text{mass})(\text{specific heat})(\text{change in temperature})$$

or

$$Q = mc\Delta T$$

The formula above is only relevant provided that a phase change does not transpire. The units for heat are either Joules or calories. A calorie is the quantity of energy required to raise the temperature of one gram of water by one degree Celsius. Body temperature is controlled when there is equilibrium between the production and the loss of heat ($Q_{\text{lost}} = Q_{\text{gained}}$). Under normal circumstances, the body dissipates heat at a rate of 100 J s^{-1}; this means that it yields nearly the same amount of heat as a 100 W light bulb.

There are three ways in which heat can be transferred: conduction, convection and radiation. Conduction involves the collision of molecules, where these collisions impact momentum and can be elastic or inelastic. A collision where kinetic energy is not conserved is called an inelastic collision. During this process, internal organs and circulating blood are at an elevated temperature. The human body naturally maintains a normal, physiological temperature of ~98.6 °F. If the body's temperature falls below its physiologic state (e.g. below 91 °F), muscle failure or even loss of consciousness may occur. Exceeding this temperature will cause the central nervous system to break down, and ultimately death will occur at ~111 °F. If the temperature of the body declines lower than its surroundings, then conduction and radiation aid in the body gaining heat. Contractions of both the liver and muscles are mainly responsible for generating heat within the body. Alternatively, the skin loses heat if the temperature of the body is greater than its surroundings. 20% of this loss is due to conduction and convection, 50% comes from radiation, and 30% is due to evaporation. **Specific heat** is 'thermal inertia', meaning that it explains the resistance of a material or object to changes in temperature. Water (c_{water} = 4.18 J g^{-1} °C^{-1} = 1 cal g^{-1} °C^{-1}) is known to have a high specific heat. It is an important molecule in the body, as it makes up about 70% of the body's mass. Various organs and tissues within the body resist changes in temperature differently. For instance, the lung has a slightly higher specific heat than the skin; yet, the average specific heat of the organs and tissues in the human body is 0.84 kcal kg^{-1} °C^{-1}. Considering, the human body is considered to have a high specific heat. Table 3.2 highlights the specific heat of different regions of the body.

Latent heat is when there is a phase change in matter and the temperature remains constant. There are two types: (1) latent heat of vaporization (L_v)—where a liquid changes to a gas; (2) latent heat of fusion (L_f)—where a solid changes to a liquid. The latent heat of fusion for water is 80 kcal kg^{-1}, while the latent heat of vaporization for water is 540 kcal kg^{-1}. An example of latent heat of vaporization is when water changes from liquid to gas through the lungs, mouth and skin. Due to evaporation, the water molecules in contact with these tissues gain enough kinetic energy to escape the liquid phase and turn into water vapor. Example 3.3 demonstrates how both specific heat and latent heat can correlate in certain scenarios.

Example 3.3. An elderly woman is walking outside of her home. She slips and breaks her ankle. The ankle results in significant swelling as shown in figure 3.1. In addition to a brace, the woman is instructed to put an ice pack on the ankle. If she places a 57 g ice pack on her ankle and it cools the area from 37 °C to 13 °C, what is the mass of the area where the ice pack is placed (note: $c_{tissues}$ = 0.84 kcal kg^{-1} °C^{-1})?

Table 3.2. Specific heat values for organs and tissues throughout the body.

Organ/tissue	Specific heat (J kg^{-1} K^{-1})	Specific heat (kcal kg^{-1} °C^{-1})
Blood	3770	0.90
Brain	3630	0.87
Epidermis	3590	0.86
*Eye	3130–4200	0.75–1.0
Fat	2350	0.56
Heart	3700	0.88
Liver	3620	0.86
Lung	3890	0.93
*Muscle	3540–3800	0.85–0.91
Stomach	3690	0.88
Body average	3500	0.84

*The specific heat for both the eye and the muscle ranges depend on the region or actual muscle.

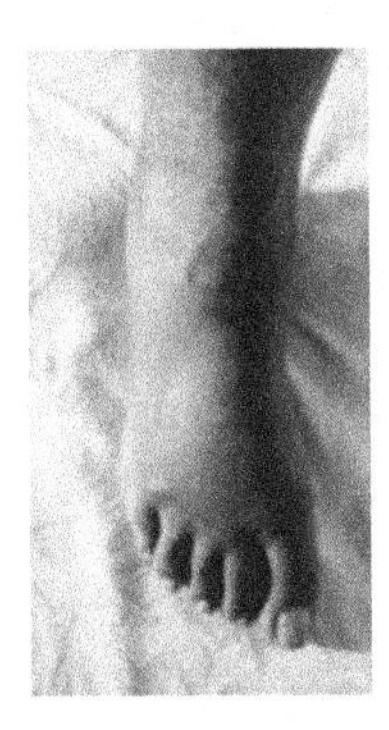

Figure 3.1. Broken, swollen ankle.

Solution: We are given the mass of the ice pack (m_{ice} = 57 g = 0.057 kg), the change in temperature (T = 13 °C − 37 °C = −24 °C), and the average specific heat of tissues is 0.84 kcal kg^{-1} °C^{-1}. An additional factor that can be added is the L_f for water, 80 kcal kg^{-1} for the ice pack. This problem is an example of the conservation of energy ($Q_{lost} = Q_{gained}$). Considering, the two formulas described above for heat will be used: $Q = mc\Delta T$ and $Q = mL$.

$$Q = m_{ice}L_f$$
$$Q = (0.057\ \text{kg})(80\ \text{kcal kg}^{-1})$$
$$Q = 4.56\ \text{kcal}$$

Next, we will use $Q = mc\Delta T$ to determine the mass of the area.

$$-Q = m\ c_{tissues}\Delta T$$
$$(-4.56\ \text{kcal}) = m(0.84\ \text{kcal kg}^{-1}\ {}^\circ\text{C}^{-1})(-24\ {}^\circ\text{C})$$
$$(-4.56\ \text{kcal}) = m(-20.16\ \text{kcal kg}^{-1})$$

0.23 kg = m**(mass of the area where the ice pack is placed)**

The **first law of thermodynamics** is a perfect explanation of how the body converts food into energy, coinciding directly to the conservation of energy. It shows that the change in internal energy equals the difference in the amount of heat added to a system and the work done by the system. The **second law of thermodynamics** states that heat will not spontaneously move, but migrates from a warm to a cold environment. This law can also be addressed on the basis of entropy (a measure of disorder); if a thermodynamic process moves from one equilibrium state to another, then the entropy of the entire system will increase or remain constant.

3.4 Metabolism

The energy that the body uses throughout the day comes from food. **Metabolism** plays an active role in regulating order within the body. It involves the net chemical reactions that take place in the body, digestion and the elimination of waste. The metabolic processes in the liver, heart, brain and skeletal muscles are the primary sources of heat. Metabolism can be catabolic—'requiring energy', or anabolic—'energy releasing'. During the catabolic state, the body processes food to use for energy; anabolism is when food is used to mend or restore cells. Catabolism is referred to as an oxidation reaction and anabolism is known as a reduction reaction. Cellular respiration is an oxygen-dependent process that is the most proficient catabolic pathway used by organisms to yield energy (~38 ATP) stored in glucose. ATP transmits chemical energy within cells for metabolism. The basal metabolic rate (BMR) is when the body converts energy into heat while at rest. This rate is slightly lower in females compared to males. If you have a male and female with a similar mass of 70 kg, the BMR for the female will be nearly 60 kcal h^{-1} while the BMR for the male would be 70 kcal h^{-1}. A higher weight results in a larger BMR. Total metabolic rate (TMR) is the overall amount of energy expended by the body to carry out all of its internal and external work. When changes in the body temperature occur, the total metabolic rate increases and more energy is burned. Nearly all of the weight in food is comprised of carbohydrates, proteins, fats and water. Fats have the highest food energy per mass of 8.8 kcal g^{-1}, while carbohydrates and proteins have a food energy per mass of 4.1 kcal g^{-1}. Most foods are composed of water, with fruits and vegetables comprising as much as 95% water. Our bodies are made up of 70% water. Water is found in three central locations in the body: within our cells, in between our cells and in our blood. It is important for proper transport of nutrients, regulating body temperature, hydration and the obliteration of waste. The threshold for life without water on average is 4 days. Polyuria is relevant to metabolism as it will occur when the body feels as though it has too much fluid and is unable to maintain it; hence, the fluid is released.

Typically, a person uses approximately 2000 calories daily. This is slightly more than the total energy that is necessary to light a 60 W bulb for 30 h. The calories used on food labels are actually kilocalories and are usually indicated by 'Calorie'—1000 calories equals 1 Calorie. For example, a 100 calorie snack contains '100 Calories"; however, to a physicist, the snack contains 100 000 calories. A person's TMR may vary depending on numerous factors including illness, age, pregnancy and depression. In general, children have a greater TMR in comparison to adults. As people age, the TMR declines. Pregnancy may increase TMR and depression can cause the TMR to decline.

3.5 Pressure

As we walk, what type of pressure and force do we apply to the ground? We learned from Newton's third law that as we exert a force on the ground, the ground responds with an equal and opposing force. Pressure is a scalar and accounts for the ratio of an applied force to a given area. It is represented as

$$\text{Pressure} = \frac{\text{Force}}{\text{Area}}$$

or

$$P = \frac{F}{A}$$

The units are Newtons per meter squared (N m^{-2}). In the SI system, the Pascal (Pa) is the unit for pressure. Additional units for pressure include: atmosphere (atm), bar, millimeter mercury (mmHg), torr, and psi (pounds per square inch). The 'mmHg' is commonly used when measuring a person's blood pressure, whereas 'psi' is common when accounting for tire pressure on an automobile or water pressure. Pressure also has a direct effect on stability and equilibrium. For example, the middle ear and the throat are connected by the Eustachian tube, which aids in equilibrating the pressure between the body and the atmosphere. At a higher pressure, it is not unusual for a person to appear to be senseless, experiencing a disturbing feeling in their ears. It is thought that if you massage the balls of your feet, it will help to drain the inner and middle ear, reducing blockage of the Eustachian tubes. Moreover, the body has 'pressure points' that align with energy pathways. Pressure points can relate to points of pain that coincide with the muscles and tendons, or reflex points that involve involuntary movements. There are over 300 pressure points that are known to help regulate function within the body. Acupuncture and reflexology are two fields that have become popular that focus on the pressure points to improve overall functional health. Considering all aspects, pressure has a way of regulating the body to ensure proper order.

Further reading

Giancoli D C 2016 *Physics: Principles with Applications* (Boston, MA: Pearson)

Hall J E and Hall M E 2020 *Guyton and Hall Textbook of Medical Physiology e-Book* (Amsterdam: Elsevier Health Sciences)

Herman I 2007 *Physics of the Human Body* (Berlin/Heidelberg: Springer)

Hewitt P 2015 *Conceptual Physics* 12th edn (Glenview, IL: Pearson)

McCall R P 2010 *Physics of the Human Body* (Baltimore, MD: Johns Hopkins University Press)

Nelson P 2004 *Biological Physics* (New York: WH Freeman) pp 315–32

Roberts T J and Azizi E 2011 Flexible mechanisms: the diverse roles of biological springs in vertebrate movement *J. Exp. Biol.* **214** 353–61

IOP Publishing

Biophysics of the Senses (Second Edition)

Tennille D Presley

Chapter 4

Electrical properties of the body

Functioning as one large electrical system, the body is driven by a variety of processes that effectively communicate. Protons, electrons and neutrons are the key components to an atom, which is the building block of matter that controls electricity. These charges facilitate electrical paths and interactions, while also orchestrating concepts of force, field, potential and magnetism. In this chapter, the importance of bioelectricity is introduced. Ohm's law is explained and each contributing concept that defines this law is described. Although power was introduced in chapter 3, its relevance to electricity and importance to the body is further expanded. In addition, magnetism, its correlation to electrical charge, and electrical energy storage is presented. Together, the information throughout the chapter provides insight into the methodologies and concepts involved in the electrical process.

4.1 Atoms and electricity

We have discussed how the body has the capacity to bend and move. Now it is important to address the inner workings of the body and how internal processes occur. All of our bodily activities are controlled by electrical signals that migrate through our bodies. These signals allow us the ability to detect movement, pain, heat, thirst and hunger, as well as how to react to these sensations. Best described as **bioelectricity**, this type of electricity impacts cell-to-cell communication and proper biological functioning for regular survival. Bioelectricity is an electric current that is produced by living tissues. Your cells are able to generate charges that 'leap' from one cell to the other, until arriving at their desired location.

The body is merely one large machine full of circuitry and electricity. Since practically 70% of the body is made up of water, it is considered to be a good conductor of electricity on average. This is due to the ions (i.e. Na^+, K^+, Cl^-) that are contained within the water. Deionized water will not conduct an electric current because it is a covalent compound, which shares electrons. An ion is something that

doi:10.1088/978-0-7503-3283-5ch4

has the ability to gain or lose an electron. Ions are important when thinking of disease as they impact the flow of charge throughout the body to maintain its functionality. For instance, the kidneys help to remove excess ions from the blood by excretion via urine and osmotic pressure regulation. In general, negative ions fortify the body's immune system and augment metabolism, whereas positive ions have the opposite effect.

At rest, your cells have an influx of potassium ions inside the cells compared to fewer sodium ions; more sodium ions are outside the cells. Because of this natural imbalance of potassium and sodium ions inside and outside of the cell, a negative resting state exists. Potassium ions are negative, while sodium ions are positive. When the body desires to send a message, the membrane gate opens and the sodium and potassium ions begin to migrate. This leads to changes in charge between the inside and outside of the cell, and an electrical impulse is triggered. An impulse is the product of force and time. Also recognized as the 'total change in momentum', impulse is most common when dealing with forces that function during a brief time interval.

Generating electricity within the body gives cells the opportunity to communicate and perform fundamental biological functions that are required for life. Cells that are not robustly transferring messages are thought to be negatively charged. The cells produce electrical charges through electrolytes (e.g. sodium and potassium). Electrolytes are solutions of ionic substances and have the capability to conduct electricity. Simply put, electrolytes are developed from ions. Many foods that have a high acidity such as lemons, potatoes, carrots and pickles, act as electrolytes and conduct electricity.

Atoms regulate electricity and are defined as the building blocks of matter. They are the smallest particle of an element and are composed of positive (protons), negative (electrons) and neutral (neutrons) charges. The basic structure of the atom is visualized as the protons and neutrons (known as nucleons) which are more massive and concentrated within the nucleus of the atom, and the electrons are assumed to circle the nucleus. This is shown in figure 4.1. Our bodies consist of mainly four atoms: carbon, hydrogen, nitrogen and oxygen. Oxygen is predominant equaling around 65%, followed by carbon (18%), hydrogen (10%), and nitrogen (3%). The majority of the atoms within the body are hydrogen; yet, oxygen makes up most of the mass. The electrons control the chemical behavior of the atom, whereas the neutrons affect the structural stability of the nucleus. The protons and electrons have a 'marriage' that exists, where they are naturally attracted to each

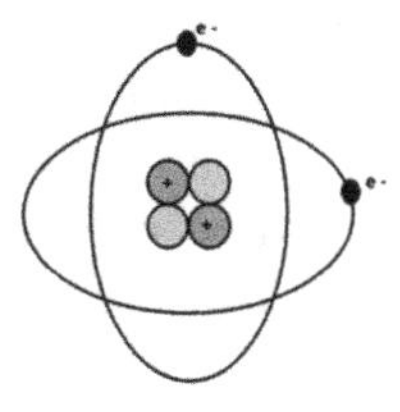

Figure 4.1. A model of the atom. Protons ('+') and neutrons are at the center; electrons (e–) are orbiting around the nucleus.

other—*'like charges repel and unlike charges attract'*. An atom is normally neutrally charged, having an equal number of protons and electrons. The charge for protons and electrons is based on 'Coulomb', which is the unit for charge. The magnitude of charge for both a proton and electron is the same, $q = 1.602 \times 10^{-19}$ Coulombs (C); since the electron has a negative charge, this value is -1.602×10^{-19} C. Harnessing the electrical charge of the Earth via grounding has been known to have a positive effect on health such as reducing inflammation, relieving pain and improving sleep. It has been suggested that the easiest way to stay 'grounded' is to walk outdoors barefoot on grass or sand. Grounding dissipates electricity.

4.2 Electric force, electric field and electric potential

Protons and electrons are just like males and females—in general, males and females like to come together to create a relationship, whether friendship or romantic. Positive and negative charges are similar. Although opposite, they attract and develop stability with each other. When the charges are stationary, there is a force that exists between these charges called an electrostatic force. This force is best represented by Coulomb's law, which states: '*The product of two charges is directly proportional to the electrostatic force and this force varies as the inverse square of the distance between the two charges*'. Thus, Coulomb's law is:

$$F = k\frac{q_1 q_2}{d^2}$$

where F is the force in Newtons, k is permittivity of free space equally to 8.99×10^9 $\text{N} \cdot \text{m}^2\ \text{C}^{-2}$, q_1 and q_2 are the charges measured in Coulombs, and d is the distance in meters. The double strands of deoxyribonucleic acid (DNA) consist of positive and negative charges, causing these strands to be attracted by electrostatic forces. The detection of the electric field is by the force that it exerts on other electric charges. It is a vector quantity that is defined as the force on a stationary charge divided by the charge and is denoted by:

$$E = \frac{F}{q}$$

The SI units for electric field can be measured as Newtons/Coulomb (N/C) or Volts/meter (V/m). In the presence of a positive charge, the charge feels a force in the direction of the electric field; however, a negative charge in an electric field experiences a force in the direction opposite to the field. Exhibiting the ability to detect electric field is electroreception. Several aquatic animals, inclusive of fish, dolphins and sharks are capable of sensing changes in the electric field in their surrounding environment. However, when it comes to humans, work is required to move an electric charge the same way that it is needed to lift a box vertically from the ground. Electric potential is merely the potential energy per charge.

Electric field and electric potential are directly proportional to each other. Therefore, a hasty change in voltage implies that there is the presence of a strong electric field. Electric field can be demonstrated by the following formula with respect to electric potential:

$$E = \frac{V}{d}$$

4.3 Current, voltage, and power

Ohm's Law demonstrates the relationship between electric potential (voltage), current and resistance ($V = IR$). Voltage is the capacity to drive an electric current across a resistance. At a fixed resistance, voltage and current are directly proportional to each other. The standard voltage in a home in the United States is 120 V, whereas it is significantly higher in other parts of the world such as Europe and Asia. The standard voltage in those continents is 240 V. Current (I) is a ratio of the amount of charge that flows past a point and the amount of time it takes the charge to flow. It is expressed as

$$I = \frac{Q}{t}$$

The unit for current is a Coulomb per second, which is equivalent to an ampere (A) or amp for short. The pressure points that were discussed in chapter 3 send electrical currents throughout the body to improve health. The majority of people can feel a current as low as 1 milliampere (mA). At higher currents between 10 and 20 mA, muscle contraction occurs. At 50 mA of current, a person feels pain. Currents that exceed this level can result in fibrillation of the heart. When there is an electric field in a conductor, charges move and electric current is created.

Resistance restricts the flow of current in an electronic circuit, and has units of the Ohm (Ω). The rate of resistance depends on the type of material that the resistor is made up of, the area and the length. Resistance is represented as

$$R = \rho\frac{L}{A}$$

where ρ is the resistivity—a property of the material that is dependent on temperature and purity. Resistance is directly proportional to length and inversely proportional to area and diameter. Resistivity is characteristic of the material, and can vary across different regions of the body, as shown in table 4.1.

Within the body, the skin (specifically the epidermis) is a poor conductor of electricity, thus implying that it has a high resistance. The skin is where most of the body's resistance exists. The resistance of the skin can range from 1000 to 100 000 Ω. This large variability depends on the skin's moisture, a person's gender and how healthy the skin may be. Dry skin can be somewhat insulating, whereas wet or blistered skin has a lower resistance. In the presence of moisture, there is less resistance and greater current. The electrical resistance varies from person to person. Yet, there are distinguishing factors between men and women. In general, men are

Table 4.1. Resistivity of some areas within the body.

Tissue	Resistivity (Ω·cm)
Blood	160
Fat	2500
Heart muscle	250 and 550 (anisotropic)
Liver	700
Lungs	2000
Skeletal muscle	150 and 2500 (anisotropic)

typically more massive and have thicker arms and legs than women. As a result, men have a lower resistance. Internal resistance in the body can vary between 300 to 1000 Ω. The route that electricity takes through the body is also important. If it enters the left hand and exits out of the right foot, then it will be considerably higher than if it goes in and out of neighboring fingers. Resistance values ranging between 500 to 1500 Ω are customary for hand-to-hand, hand-to-foot and hand-to-foot.

Alternating current (AC) is when the current frequently changes direction, while direct current (DC) is a consistent flow of current in one direction. AC is most common of the two, with respect to its utilization in buildings, homes, the body, and practically daily life. Each of their effects can impact the body in different ways. Both alternating current (AC) and direct current (DC) can be fatal; however, the path in which the current travels through the body is a major factor. DC will create a single, continuous contraction of muscles, whereas AC will generate a series of contractions (*depending on the frequency*). If current travels across the body, and through the heart, lethal effects can result and possibly death.

Power was discussed in chapter 3 as the rate of energy transformation. As this concept relates to electrical activity, **power** is the product of current and voltage ($P = IV$), and still has the units of Watts. If we correlate Ohm's law, two additional formulas for power can be resolved:

$$\text{Power} = (\text{Current})^2(\text{Resistance})$$

$$\text{Power} = \frac{\text{Voltage}^2}{\text{Resistance}}$$

While $P = IV$ can be useful for any electrical device, the two equations above can only be used if a resistance is present. There is a direct correlation with how power is viewed with respect to energy compared to electricity. For instance, the electric companies measure more of energy that a home uses and not the power. In a household circuit, a fuse or circuit breaker opens when there is an overload of current. The body has a similar pattern where it will regulate current to ensure that there is not too little or too much current. A human body can produce between 10 and 100 millivolts. A large voltage simply means that there is a high potential for current to rapidly flow through the body. If you are taking a shower or in the pool, I suspect that you avoid using your cell phone or any other electrical device.

As people, we have a higher conductivity when wet, increasing the likelihood of electric shock or electrocution. Electric shock disturbs the normal operations of the body. An extreme shock such as lightning can be the result of electrocution and of course the body's resistance is drastically depleted. A person's body chemistry can also have a substantial effect on electric current.

The dominant ions within the body were discussed, and each ion has an ion channel. An ion channel is responsible for regulating the flow of ions across the cell membrane. The electrical properties of biological tissues and cells can be studied via electrophysiology. This method entails measurements of voltage change or electric current from single ion channel proteins to large organs. Patch clamping is a common technique to study a single channel or the entire electrical activity of a cell. With patch clamping, a glass micropipette is filled with an electrolyte solution and the tip of the micropipette is attached to the cell membrane.

4.4 Magnetism

While electric charges can generate an electric field when stationary, magnetic fields require 'movement' whether it is a result of charges that are migrating or current in wires. These ions move when subjected to an electric field. When moving, these charges have the capability to interact with magnetism. A magnetic force is exerted on a charge that moves in a magnetic field. This force can be denoted by:

$$F = qvB \sin \theta$$

As demonstrated by the formula, the magnetic force is also influenced by the velocity of the charge and by the magnetic field (B), having units of Tesla (T). *Theta* (θ) is the angle between the velocity and the magnetic field. A changing electric field can cause a change in the magnetic field and a changing magnetic field can lead to a variation in the electric field. This is best explained as electromagnetism, as electricity and magnetism have a direct effect on each other. The interaction of electric and magnetic fields is a perpendicular relationship with each other. It is possible to detect the direction that someone is facing based on the Earth's magnetic field, known as magnetoception. The Earth has a magnetic field of 5.0×10^{-5} T. While this sense is not common amongst humans, it is observed in birds and bees.

The human body generates an electromagnetic field, causing the forces of electricity and magnetism to coexist. The electrical portion consists of a low frequency, DC electric field. When these forces are present, electrical energy also exists. This electromagnetic energy is impacted by the natural electromagnetism of the earth. Although the earth's electromagnetic environment is typically unobtrusive, it may become disturbed by an environment that may be electromagnetically charged. The overlap of electricity and magnetism within the biological system can also be studied via electrophysiology.

4.5 Capacitance

A capacitor is a set of conducting plates that has charges of equal magnitude, but opposite signs. Capacitance is a measure of a capacitor's ability to store electric potential energy or charge. This storage of energy is equivalent to the work done to charge the capacitor, and occurs by disassociating positive and negative charges. The charges from one plate are transferred to the other plate, as commonly modeled by a battery that is conjoined to a capacitor. Capacitance is represented by the ratio of charge to voltage and can be determined by the formula:

$$\text{Capacitance} = \frac{\text{Charge}}{\text{Voltage}}$$

The unit for capacitance is a Coulomb per Volt which is a Farad (F). Similar to any other electrically conductive material, the human body can store electric charge if properly insulated. The capacitance for the human body in a normal surrounding is around 10–105 picofarads (pF). While humans are larger than many electronic devices, this capacitance is small compared to electronic standards and people usually are well-spaced from other conductive objects. The attire that you wear will impact your capacitance. Direct contact with a conductor reduces a person's capacitance, whereas an insulating material will enhance the capacitance. A person's posture also influences their capacitance.

Dielectrics are simply insulating materials. The dielectric constant for human tissue ranges between 30 and 80. Tissues from the lungs have a lower dielectric constant than liver and muscle tissues. In a typical electrical circuit, a battery (or source of power) is present with a positive charge on one end and a negative charge on the other end. When you connect the battery within a complete circuit, there is usually some type of grounding, which prevents too much current from flowing through the circuit. The purpose of a battery is to generate a difference in potential and trigger the movement of charges. In order for current to flow, a complete circuit must be present. However with the human body, charges on either side of the cell membrane are created through passive processes such as simple diffusion, migration through protein channels in the membrane or active-energy dependent processes involving ATP. With respect to a dielectric, capacitance is represented as

$$\text{Capacitance} = K\epsilon_0\frac{A}{d}$$

where K denotes the dielectric constant. This value varies depending on the particular material that is present. ϵ_0 is a permittivity constant equivalent to $8.85 \times 10^{-12}\ \mathrm{C^2\ N^{-1} \cdot m^{-2}}$; A represents the area and relies on the shape of an object and d is the displacement. Capacitance is dependent on physical characteristics, and not the charge or voltage.

We currently live in an era where technology significantly impacts our lives. Have you ever questioned how is it that when you touch your smartphone or tablet, the screen recognizes your touch? Your finger has a different dielectric constant compared to the air. It changes the electric field and the mutual capacitance of the wires. When you place your finger near the screen, it senses the capacitance between the electrode or wire and your body. This is possible due to small capacitor grids that team up so that one direction of wires carries current and the other senses the capacitance between them. This is a direct correlation of the bioelectricity in the body and how you transfer charges from your body to the screen.

It is appropriate to consider the energy stored in a capacitor as being deposited in the electric field between the plates of the capacitor. When there is a lot of energy accumulated in a large capacitor, it can be detrimental and cause a burn or electric shock if you come into contact with the capacitor. Capacitors can carry charge even if the power source is deactivated. Thus, be extremely careful in the presence of circuits. Electric potential energy stored in a capacitor is

$$\text{Potential energy} = \tfrac{1}{2}\,QV = \tfrac{1}{2}\,CV^2 = \tfrac{1}{2}\frac{Q^2}{C}$$

Example 4.1 If two regions of fat ($K = 16$) are 3.0 cm apart and experience 60 V, how many joules are stored between them if their area is 8.0 cm?

Solution: (a) First, we need to identify what we are given in the problem: $K = 16$ for fat, $d = 3.0$ cm = 0.03 m, $V = 60$ V, $A = 8.0$ cm = 0.08 m. Thus, we must first determine capacitance:

$$\begin{aligned}\text{Capacitance} &= K\epsilon_0\frac{A}{d}\\ &= (16)(8.85 \times 10^{-12}\ \text{C}^2/\text{N}\cdot\text{m}^2)\left(\frac{0.08\ \text{m}}{0.03\ \text{m}}\right)\\ &= 3.77 \times 10^{-10}\ \text{F}\end{aligned}$$

To determine the potential energy (Joules), we use

$$\begin{aligned}\text{Potential energy} &= \tfrac{1}{2}\,CV^2\\ &= \tfrac{1}{2}(3.77 \times 10^{-10}\ \text{F})(60\ \text{V})^2\\ &= \mathbf{1.36 \times 10^{-6}\ J\ or\ 1.36\ \mu J}\end{aligned}$$

When positioned in contact with a piece of metal, the epidermis will function as a capacitor. The tissue directly beneath the epidermis and the metal act as the conductive plates of a capacitor; the dry surface of the epidermis can act as the dielectric. Static shock has low amounts of energy stored and is not harmful. A high frequency electric current or AC is far more dangerous than a DC current. Both AC and DC can be fatal; however, it takes more milliamperes of DC than AC at the same voltage to have the same effect. If electric shock occurs via AC, the body's overall resistance is significantly lowered; however, electric shock by a DC source will not have as great an effect. For instance, an electrical shock of 1 mA of AC at 60 hertz would be equivalent to 4 mA of DC. The probability of severe shock enhances with a greater amount of voltage. A voltage exceeding 450 V causes dielectric breakdown of the skin. Electric shock that results in death is considered as electrocution.

Think about a solar-powered watch. If you place the watch in the Sun to let it charge and then place it on your arm, the body's electrical activity will interact with the normal workings of the watch to keep the watch functioning to tell time. There are also watches that are fueled by the conversion of kinetic energy from the body to electrical energy. There has also been research exploring the body as a 'human battery'. For example, Panasonic has developed a method for human blood to be used to 'power' electrical devices. The intent is that this could aid with nanodevices that may be implanted in the body. From minute changes in temperature to harnessing kinetic energy, the body can be useful for low powered electronic devices of 100 microwatts.

Reflect on how you feel at the end of a very long day. If you are anything like me, you feel exhausted at the end of the day. The goal is usually to have dinner, relax and rejuvenate for the next day. Your body is a natural capacitor in the fact that food and sleep aid in providing the body with the right amount of energy so that it can 'recharge' and be prepared for the next day. All-in-all, there are so many useful ways that the electricity from the body is important.

Further reading

Becker R O 1990 *Cross Currents: The Perils of Electropollution, the Promise of Electromedicine* (Los Angeles, CA: Jeremy P. Tarcher)

Becker R O, Selden G and Bichell D 1985 *The Body Electric: Electromagnetism and the Foundation of Life* (New York: Quill)

Davis L 2019 *Body Physics: Motion to Metabolism* (Corvallis, OR: Oregon State University)

Fujiwara O and Ikawa T 2002 Numerical calculation of human-body capacitance by surface charge method *Electron. Commun. Jpn.* **85** 38–44

Giancoli D C 2016 *Physics: Principles with Applications* (Boston, MA: Pearson)

Hewitt P 2015 *Conceptual Physics* 12th edn (Glenview, IL: Pearson)

Rush S, Abildskov J A and McFee R 1963 Resistivity of body tissues at low frequencies *Circ. Res.* **12** 40–50

Schwan H P and Kay C F 1957 Capacitive properties of body tissues *Circ. Res.* **5** 439–43

Weyer S, Ulbrich M and Leonhardt S 2013 A model-based approach for analysis of intracellular resistance variations due to body posture on bioimpedance measurements *J. Phys. Conf. Ser.* **434** 012003

IOP Publishing

Tennille D Presley

Chapter 5

Free radicals

'Free' means *to set loose from restraint*, while 'radical' is considered to be *different from the ordinary*. When combining these two terms together, a 'free radical' is when something out of the ordinary is released. This is the true essence of a free radical. Free radicals are everywhere throughout the environment, making it very difficult at times to live with them and to live without them—they are essential. Understanding the 'good' radicals versus the 'bad' is key, as well as the physiologic concentrations that are beneficial in comparison to being harmful are important. Free radicals tend to fall into two categories: reactive oxygen species (ROS) and reactive nitrogen species (RNS). In this chapter, free radicals are introduced and described based on each of their unique characteristics. Antioxidants are also discussed, which are ways to modulate radicals. Lastly, measurements and techniques that can be performed to determine free radical concentrations are described.

5.1 What is a free radical?

If you ruled the world, what would you do or what would you be? You can ask yourself the question of what would you do; however, the answer to what you would be is a free radical. A **free radical** is a chemically reactive atom, molecule or ion that has an unpaired electron. Free radicals are thought to have 'dangling' covalent bonds, and in a sense rule the world because they are everywhere. Free radicals usually arise during metabolism, but can also be generated by cigarette smoke, pollution and radiation. It is impossible to consider an environment lacking these radicals. When there are more free radicals, there are more positively charged ions. If there are more free radicals in the blood, cellular metabolism is reduced and less efficient. As a result, the cells are weakened and the body is more prone to sickness and aging. This may also initiate failure to the body's immune system. When you are sick, you have a higher amount of free radicals. Think about how you feel. In most cases, you feel depleted of energy. This let's you know that the free radicals are

doi:10.1088/978-0-7503-3283-5ch5

heavily present. Free radicals are 'robbers' of energy. They wait for the most opportune time, then they take over and impact our daily lives (figure 5.1).

Free radical damage exists within our bodies and our surrounding environment. This damage transpires when an unpaired electron seeks another electron that is paired to a molecule. The unpaired electron normally approaches a weak bond and 'steals' the electron, causing the 'attacked' molecule to become a radical; thus, a chain reaction and disruption of a living cell occurs. Stress or any type of imbalance instigate the prevalence of free radicals. Diet, exercise, disease, and aging also influence radicals. The free radical theory of aging states '*Organisms age because cells accumulate free radical damage over time*'. Damage from free radicals within the cells is associated with a number of diseases including cancer, diabetes, Alzheimer's diease, cardiovascular disease and stroke. However, we cannot survive without free radicals, as they are also involved in vital processes within the body to regulate homeostasis. For instance, free radicals aid in abolishing bacteria.

5.2 Types of free radicals

The most problematic free radicals within biological systems are the oxygen radicals. Since oxygen has two unpaired electrons in its outer shell (in separate orbitals), it has a high probability of generating radicals. Known as reactive oxygen species (ROS), these oxygen-centered radicals include the hydroxyl radical, superoxide and hydrogen peroxide. They are constructed from normal oxygen metabolism, and impact homeostasis and cell signaling; their half-lives are demonstrated in table 5.1. ROS are involved in various enzymatic reactions, and are also reduced along the electron transport chain; although during oxidative stress, ROS may markedly increase. This effect can be triggered by overexposure to heat, ionizing radiation, abnormal concentrations of oxygen or physical exhaustion. A perfect example of an overproduction of oxygen radicals is with white blood cells. These cells tend to produce oxygen radicals to kill invading pathogens. Ionizing radiation is a common producer of ROS within biological systems, and is most detrimental in highly oxygenated tissues. Long-term consequences of ionizing radiation is DNA damage. Numerous daily activities have also been shown to enhance the production of ROS inclusive of exposure to sunlight, cigarette smoke, and air pollution. Sun exposure has been shown to encourage oxidative destruction of the skin, augment the probability of skin cancer and induce wrinkling of the skin.

Singlet oxygen ($^1O_2$), an excited form of oxygen, is another oxygen-derived radical. With $^1O_2$, one of the unpaired electrons hops to a superior orbital after

Figure 5.1. Free radicals are 'thieves'. They steal energy, leaving the body with disorder.

Table 5.1. List of common free radicals and their half-lives.

Reactive species	Half-life
Hydrogen peroxide	minutes
Nitric oxide	seconds
Peroxynitrite	milliseconds
Superoxide anion	microseconds
Hydroxyl radical	nanoseconds

energy absorption. While singlet oxygen is a prooxidant, the main ROS are superoxide, hydrogen peroxide and the hydroxyl radical. Also known as hyperoxide, superoxide (O_2^-) is generated when oxygen is reduced by a single electron; molecular oxygen gains an electron. Superoxide has a molecular weight of 32 g mol^{-1}, is paramagnetic and occurs widely in nature. A great source of superoxide is NADPH oxidase. It becomes activated by inflammation and vasoactive factors. O_2^- may also be produced during the oxidation of hemoglobin to methemoglobin. An increase in superoxide is believed to enhance the aging process. Superoxide and hydroxyl are generally associated with cytotoxicity.

Hydrogen peroxide (H_2O_2) is a strong oxidizer that is naturally produced by oxidative metabolism. It is the most significant ROS in cell cycle regulation, and important for the immune system. H_2O_2 is most recognized as a common solution found in a person's medicine cabinet. H_2O_2 is commonly used to disinfect wounds, whiten teeth, treat acne, and for hair bleaching. Hydrogen peroxide expedites wound healing. Through catalysis, hydrogen peroxide can be rapidly transformed into hydroxyl radicals. Hydrogen peroxide is also recognized for its detrimental effects. For instance, a person with asthma has higher levels of hydrogen peroxide in their lungs, leading to unsuitable amounts of white blood cells and moments of difficulties in breathing.

The hydroxyl ($OH^•$) radical is recognized for its popular characteristics of extreme toxicity and reactivity. $OH^•$ can lead to the formation of hydrogen and oxygen. It usually attacks everything that it collides with, especially phospholipids in cell membranes and proteins. Unlike superoxide, the hydroxyl radical cannot be eradicated by an enzymatic reaction.

Another category of reactive species are the reactive nitrogen species (RNS). RNS are 'a family of nitrogen molecular entities that are derived from nitric oxide ($NO^•$) and O_2^-'. Discovered as the endothelium-derived relaxation factor (EDRF), $NO^•$ or nitrogen monoxide is a stable radical that is naturally produced in the body. In the blood, NO has a very short half-life of just a few seconds, which is highlighted in table 5.1. It is present in the body both endogeneously and exogeneously. At low amounts, NO production is protective against ischemic damage in organs such as the liver. Nitric oxide is best known for its vasodilation. It is also commonly recognized to inhibit platelet aggregation, protects against vascular dysfunction, and exhibits properties of a neurotransmitter. $NO^•$ aids in processing nerve signals as they cross synapses. Out of the 20 amino acids that comprise proteins, L-arginine is the only one that makes substantial amounts of $NO^{\bullet}$. Limited amounts of NO contribute to endothelial dysfunction; the gut is a key site for high concentrations of NO.

NO is an efficient intracellular messenger because of its ability to diffuse through most cells and tissues, causing very little disturbance. NO impacts many diseases, especially those that affect the vasculature. This signaling molecule initiates homeostatic functions including cell-to-cell communication, wound healing, pain reduction, cell proliferation, and the immune response. The effect of nitric oxide on neurotransmission happens by cyclic guanosine monophosphate (cGMP) allowing the phosphorylation of ion channels. Adequate levels of oxygen and glucose are carried to the nerve cells, improving the production of ATP. NO also increases cGMP, and uses its vasodilating properties to reduce pressure on nerves; this results in a decreased level of pain. Inhaled NO has been a useful treatment for pulmonary hypertension.

$NO^{\bullet}$ can be studied from both an endogenous and exogenous perspective. Endogenously, nitric oxide synthase (NOS) is an enzyme that creates $NO^{\bullet}$ from L-arginine. There are three isoforms of NOS: NOS1 (neuronal NOS, nNOS); NOS2 (inducible NOS, iNOS); and NOS3 (endothelial NOS, eNOS). nNOS aids in transmitting information between the nerves and the brain. iNOS is named for its need to become activated when an abnormality (i.e. injury, disease, etc) exists. Extremely high concentrations of NO (100–1000 times the normal amount) are produced by iNOS. This isoform has been observed in the brains of individuals with multiple sclerosis.

Amongst the three isoforms, eNOS is important in supporting normal activity in the blood vessels. eNOS is found in the inner lining of blood vessels (the endothelium) and enhances the growth of new blood vessels (called angiogenesis). The activation of eNOS happens by the stretching and relaxation of the blood vessel wall in response to each heartbeat. eNOS is also known to play a role in affecting cellular respiration. This isoform is known to associate with various proteins such as heat shock proteins (Hsps). Hsps are proteins that are present in all cells at every biological level and aid in cardioprotection. Specifically, eNOS can conjoin to Hsp90 to enhance NO production, inhibit the generation of superoxide and regulate vascular tone. NO generated from eNOS preserves the diameter of the blood vessels to maintain adequate blood flow throughout the body.

Exogenously, nitrite (NO_2^-) and nitrate (NO_3^-) contribute to the availability of NO. Nitrate is reduced to nitrite, which is then reduced to NO. Low concentrations of NO_2^- in tissues range between 1 and 20 μM, whereas nanomolar concentrations of 100–200 nM exist in the blood. NO_2^- therapeutics has grown over the years, where it aids to improve blood pressure and the presence of NO, specifically in conditions of low availability of NO. Two primary sources of nitrate and nitrite are by the endogenous L-arginine–NO pathway and through the diet. Most NO_2^- comes from the oxidation of 'NOS-generated NO'. A reaction of NO with oxyhemoglobin produces nitrate, while oxidation of NO forms nitrite. During hypoxia, NO produced by NOS is limited, but the nitrate–nitrite–NO pathway is amplified. Beet juice and green, leafy vegetables such as spinach are also a great sources of nitrates.

Melanin is unique in the fact that it is not chemically reactive like other radicals. It is normally found in high concentrations in individuals of African descent. Melanin has 'superconducting' capabilities as it can conduct electricity and also has insulating properties. It will not allow electrical current to pass through its structure. Melanin is also protective of cells against oxygen toxicity, and aids with tissue repair and skin regeneration. It is present in the nervous system, the blood, brain and a number of foods.

5.3 Interrelationships of radicals

Knowing that free radicals are highly reactive, it only makes sense that they have the ability to react with each other. Yet, what happens and how do these processes occur? Recent reports have shown that one of the main causes of heart failure is the NO/$ONOO^-$ cycle. Nitric oxide and superoxide generally have an inversely proportional relationship with each other, where the concentration of one radical increases, the other will decrease. There are exceptions to this where increases in NO in contracting skeletal muscle of aged mice does not result in a decline in superoxide. When superoxide and nitric oxide coexist, they chemically react to form of peroxynitrite ($ONOO^-$) and cause NO toxicity. This reaction causes NO toxicity, and may compete with the dismutation of superoxide and hydrogen peroxide. $ONOO^-$ can be damaging to both DNA and proteins with cells. It can lead to lipid peroxidation and inactivation of enzymes. Peroxynitrite is also capable of forming the hydroxyl radical and nitrogen dioxide. Lipid peroxidation is a common example, which is initiated when a free radical removes a hydrogen atom from a polysaturated fatty acid.

An oxidation-reduction (redox) reaction is when electrons are moved from one atom to another. Oxidation is when there is a loss of electrons by an atom and reduction is when the atom gains electrons. Thus, any reaction that has oxidation must also have reduction—occurrence of a redox reaction. The 'redox' properties of iron are a source of free radicals in biological systems. A reversible redox reaction also exists with nitrite and nitric oxide. This interplay impacts brain function, the gut and the digestive system.

5.4 Antioxidants

As a child, I distinctly remember my parents and grandparents saying to eat my vegetables. It was not until I became an adult that I realized this reason was to combat free radicals by way of antioxidants. An antioxidant is a stable molecule that has the ability to donate an electron to neutralize the free radical—they scavenge free radicals. Antioxidants are reducing agents that inhibit oxidative damage to biological systems—they neutralize free radicals. A balance between free radicals and antioxidants is essential for normal physiological stability and function. Most antioxidants are either derived from the diet or enzymes within the body. β-Carotene, vitamin C (ascorbic acid), and vitamin E (α-tocopherol) are from the diet. Vitamins C and E can also be used topically to reverse damage from the Sun. Vitamin E is a fat soluble vitamin that is present in whole grains, nuts, vegetables and fish oils.

Vitamin C or ascorbic acid is a water soluble vitamin that is found in citrus fruits, broccoli, cabbage, green peppers, kiwi and strawberries. It is essential for tissue collagen production and protective against plaque formation in the arteries. Although vitamin A is not an antioxidant, its precursor, beta-carotene is. Beta carotene is present in milk, carrots, yams, peaches, spinach and egg yolk. Green tea is an excellent source of catechins, known to augment the immune system and cardiac health. In general, most people do not naturally obtain the necessary amount of antioxidants daily. Specifically with vitamins C and E, the American diet is low in the recommended amounts. On average, an adult should consume approximately 15 milligrams of vitamin E per day.

In the absence of catalase, toxicity from H_2O_2 transpires. As previously discussed, free radical chain reactions may occur. Due to the high toxicity and reactivity of superoxide, it is imperative that all organisms in the presence of oxygen have a protective enzyme to neutralize superoxide. Superoxide dismutase (SOD) is capable of rapidly counterbalancing the detrimental effect of superoxide, into hydrogen peroxide and oxygen. This class of enzymes comes in three forms: SOD1, which is located in the cytoplasm; SOD2, located in the mitochondria; and SOD3, found extracellularly. SOD1 and SOD3 contain copper and zinc, while SOD2 has manganese in its reactive center. Glutathione is the only antioxidant capable of destroying the hydroxyl radical.

5.5 Ways to measure free radicals

Due to the short half-lives of most radicals, scientists and researchers have developed unique ways to detect the presence of these radicals in biological tissues and samples. The most common approach for measuring free radicals is to utilize **electron paramagnetic resonance** (EPR). This is a technique that measures energy absorption due to the transition of an atomic particle between energy levels. During EPR, there is a fixed frequency in the presence of an electromagnetic field. The unpaired electrons interact with a magnetic field. EPR is based on the following equation:

$$E = hf$$

where E is the energy absorbed (in Joules), h is Planck's constant (6.63×10^{-34} J·s), and f is the frequency (often measured in gigahertz or megahertz). Free radicals in cellular samples, intact tissues, organs and whole animals can be measured. Since EPR has a fixed frequency, the type of EPR machine will vary depending on the type of sample that is used for measurement. For example, free radicals within cellular samples are generally measured using an X-band EPR, whereas whole animal (e.g. mice) measurements use L-band EPR (figure 5.2). This information is displayed in table 5.2.

EPR can be spectroscopic or via imaging, where a specific paramagnetic probe is necessary for detecting nitric oxide, superoxide and the hydroxyl radical. Because each of these radicals are 'short-lived', a spin trap is necessary for EPR

Figure 5.2. Electron paramagnetic resonance spectrometer. This is an example of an X-band EPR machine. In the center is the cavity where the sample is placed. The cavity can vary depending upon the type of EPR, as well as the sample that is being measured.

Table 5.2. Types of EPR with the corresponding microwave frequency.

Microwave band	Microwave frequency (GHz)
L	1.1
S	3.0
X	9.7
K	24.0
Q	35.0
W	94.0

measurements. When a spin trap interacts with one of these radicals, it forms a covalent bond to 'stabilize' the radical so that it can be detected by EPR. A metastable radical is created with a half-life of 1–15 min. Examples of spin traps include 5,5-dimethyl-1-pyrroline-N-oxide (DMPO), which detects superoxide, and *N*-methyl-D-glucamine dithiocarbamate iron (FeMGD), which measures nitric oxide. In some studies, the desire to match the temperature condition of the EPR experiment to the actual environment (i.e. free radical formation in biological tissues) exists. To measure cellular activity, room temperature EPR measurements can be performed; however, low-temperature EPR is for reactions that take place in the body rapidly. For example, the reaction of hemoglobin and nitrite within whole blood must be measured at a low temperature.

Other techniques exist that can measure the concentration of free radicals in biological samples and tissues. The ENO-20 is a high performance liquid chromatography (HPLC)-based technique that measures both nitrite and nitrate. It is a highly sensitive technique that can perform measurements as quickly as five minutes, with a microliter volume of sample. The superoxide dismutase (SOD) assay uses tetrazolium salt to detect superoxide in plasma, tissues and cells. It can detect all three types of SOD (Cu/Zn, Mn and FeSOD). Fluorescent probes are also commercially available to measure radicals. 5-Diaminofluorescein diacetate (DAF-2DA) is commonly used to detect the presence of nitric oxide, while dichlorofluorescin diacetate (DCFDA) measures ROS. NO can also be measured with a Clark electrode. This electrode is a specific electrochemical senso that causes NO to diffuse through a gas-permeable membrane and a thin film of electrolyte.

Oxidation occurs on the electrode, generating a current directly proportional to the concentration of NO on the outside of the membrane.

Whether it is paint peeling on our homes, pollutants in the environment, or reactions within the body, free radicals impart themselves into our lives. The key is to have a thorough understand of the benefits and dangers of all radicals, and counterbalance the harmful effects with antioxidants. Yet, remember that too much or too little of anything can have detrimental effects—everything in moderation.

Further reading

Bagasra O, Michaels F H, Zheng Y M, Bobroski L E, Spitsin S V, Fu Z F, Tawadros R and Koprowski H 1995 Activation of the inducible form of nitric oxide synthase in the brains of patients with multiple sclerosis *Proc. Natl Acad. Sci. USA* **92** 12041–5

Bahrami B, Lavie N and Rees G 2007 Attentional load modulates responses of human primary visual cortex to invisible stimuli *Curr. Biol.* **17** 509–13

Beckman J S and Koppenol W H 1996 Nitric oxide, superoxide, and peroxynitrite: the good, the bad, and ugly *Am. J. Physiol.-Cell Physiol.* **271** C1424–37

Berry M J, Justus N W, Hauser J I, Case A H, Helms C C, Basu S, Rogers Z, Lewis M T and Miller G D 2015 Dietary nitrate supplementation improves exercise performance and decreases blood pressure in COPD patients *Nitric Oxide* **48** 22–30

Jones A M 2014 Dietary nitrate supplementation and exercise performance *Sports Med.* **44** S35–45

Lundberg J O, Weitzberg E and Gladwin M T 2008 The nitrate–nitrite–nitric oxide pathway in physiology and therapeutics *Nat. Rev. Drug Discov.* **7** 156–67

Miller G D, Marsh A P, Dove R W, Beavers D, Presley T, Helms C, Bechtold E, King S B and Kim-Shapiro D 2012 Plasma nitrate and nitrite are increased by a high-nitrate supplement but not by high-nitrate foods in older adults *Nutr. Res.* **32** 160–8

Pearson T, McArdle A and Jackson M J 2015 Nitric oxide availability is increased in contracting skeletal muscle from aged mice, but does not differentially decrease muscle superoxide *Free Radical Biol. Med.* **78** 82–8

Pereira C, Ferreira N R, Rocha B S, Barbosa R M and Laranjinha J 2013 The redox interplay between nitrite and nitric oxide: From the gut to the brain *Redox Biol.* **1** 276–84

Rosen G M, Britigan B E, Halpern H J and Pou S 1999 *Free Radicals: Biology and Detection by Spinn Trapping* (Oxford: Oxford University Press)

Salisbury D and Bronas U 2015 Reactive oxygen and nitrogen species: impact on endothelial dysfunction *Nurs. Res.* **64** 53–66

Shiva S 2013 Nitrite: a physiological store of nitric oxide and modulator of mitochondrial function *Redox Biol.* **1** 40–4

www.eicom-usa.com

www.caymanchemical.com

Chapter 6

Cardiac conductivity

Whether it is the heart or circulating blood, the body cannot effectively utilize any of its senses without these two components. When considering the heart, love is one of the primary emotions that comes to mind; yet, recognizing the importance of the inner workings of the heart is key. The conductivity within the body is influential in activating cell-to-cell communication and creating currents within the myocardium. How this conducting system works, entails intricate processes that facilitate life. In this chapter, the complexities of the heart and blood flow are introduced. The impact of membrane potential on the cardiac system is also addressed. Both the electrocardiogram and echocardiogram provide mechanisms to monitor the electrical activity and sounds of the heart, respectively. When the electrical activities malfunction, cardiac abnormalities occur. Levels of dysfunction can exist in a variety of ways, which are described towards the end of the chapter.

6.1 Machinery of the heart and blood flow

The heart is the dominant organ within the body's circulatory system that is responsible for pumping blood throughout the body. It exhibits perfect machinery to keep the body functioning in the way that it needs to survive, by contracting at regular intervals to squeeze blood into the blood vessels. It contracts and relaxes due to the flow of electrical activity. This electrical conduction system creates impulses along specialized pathways, having a distinct sequence for contraction. The electrical activity of the heart also relies on electrolytes to properly function. The heart consists of four chambers: the upper chambers—right atrium and left atrium; the lower chambers—left ventricle and right ventricle. The atria and ventricles are electrically separated from each other by connective tissue, which acts like an insulator. The main controller of the normal heart rhythms is the sinoatrial (SA) node, which is located in the right atrium. The SA node produces impulses that stimulate the cardiac muscle or the myocardium, causing contraction of the myocardium. Having a rate of 40–60 beats per minute (bpm), the atrioventricular (AV) node is

doi:10.1088/978-0-7503-3283-5ch6

also instrumental in cardiac electrical activity. It attenuates the conduction of electrical impulses from the atria, and may function as a 'backup' to the SA node. This well-organized process allows blood to be pumped throughout the body.

Systemic circulation involves oxygenated blood being carried away from the heart to the body, and returning deoxygenated blood back to the heart. The right chambers of the heart contain deoxygenated blood and the left chambers house the oxygenated blood, as shown in figure 6.1. The deoxygenated blood is pumped into the lungs for reoxygenation; the left chambers receive the oxygenated blood from the lungs and the blood is pumped throughout the body to provide nourishment for the tissues and organs within the body. Hemoglobin is a protein molecule in the blood (specifically the red blood cells) that transports oxygen from the lungs to the body's tissues, and returns carbon dioxide from the tissues to the lungs. It releases oxygen and provides energy to aid in metabolism. Negative ions generate a rise in the affinity for hemoglobin, causing the partial pressure of oxygen to augment and the partial carbon dioxide pressure to decline. Due to the fact that blood is forced out with a greater resistance from the left ventricle, it is thicker than the right ventricle. However as a whole, very little effort is needed to fill the ventricles, because gravity plays a role. The atria contract before the ventricles. Blood exits the left ventricle and migrates to the aorta, the largest artery in the body. Blood flows through the aorta at a velocity of 30 cm s^{-1}. It travels through various arteries, capillaries and veins and re-enters the heart via the right atrium.

Responsible for pumping blood to the body, the heart is impacted by pressure. Pressure is the amount of force per unit area. This concept is also recognized as the product of the density (ρ) of an object, the acceleration due to gravity (9.8 m s^{-2} on earth), and the height (h) of the object. Density is mass per volume. Pressure can be determined using the following formulas:

$$\text{Pressure} = \frac{\text{Force}}{\text{Area}}$$

$$P = \frac{F}{A}$$

$$\text{Pressure} = (\text{density}) \times (\text{acceleration due to gravity}) \times (\text{height})$$

$$P = \rho g h$$

The density of whole blood is 1.05×10^3 kg m^{-3}. Whole blood represents about 7% of the body's weight. Blood pressure (also known as arterial blood pressure) is the pressure that circulating blood applies to the walls of the blood vessels. The amount

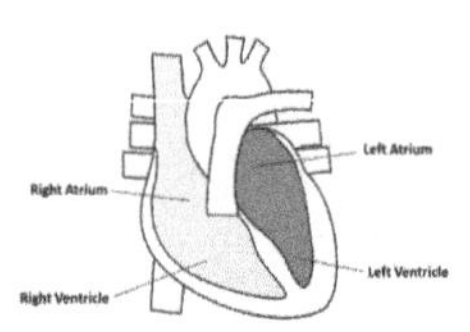

Figure 6.1. Model of the heart. The heart consists of four chambers: the right atrium, left atrium, right ventricle, and the left ventricle. The right chambers of the heart house the deoxygenated blood (shown in blue); the left chambers of the heart contain the oxygenated blood (shown in red).

of force that the blood exerts on the vessel wall impacts pressure, which is highlighted in Example 6.1. Systolic blood pressure (the top number) is the amount of pressure that the blood exerts on the vessels while the heart is beating; diastolic blood pressure (the bottom number) is the pressure in the vessels in between heartbeats. Blood pressure is read as the systolic pressure over the diastolic pressure. A normal blood pressure at a relaxed state is 120/80 millimeters of mercury (mmHg), and a standard resting heart beat is between 60 and 100 beats per minute (bpm).

Example 6.1. How much force is applied to the blood vessel ($A = 4.91 \times 10^{-4}$ m^2) wall as blood flows from the top of the head to the feet of Donald,who is 75 inches tall and standing vertically?

Solution: First, we should identify the given pieces of information:

the area (A) = 4.91×10^{-4} m^2
blood → density (ρ) = 1.05×10^3 kg m^{-3} (for whole blood)
height (h) = 75 inches = 1.91 m (*refer back to chapter 1 for the conversion)
vertically → $g = 9.8$ m s^{-2}
force = ?? (unknown)

Based on what we are given,we can determine the pressure using

$$\begin{aligned} P &= \rho g h \\ &= (1.05 \times 10^3 \text{ kg m}^{-3})(9.8 \text{ m s}^{-2})(1.91 \text{ m}) \\ &= 19\,653.9 \text{ N m}^{-2} \end{aligned}$$

Next, plug in the values for pressure and area using the equation below

$$P = \frac{F}{A}$$

$$19\,653.9 \text{ N m}^{-2} = \frac{F}{4.91 \times 10^{-4} \text{ m}^2}$$

$$9.65 \text{ N} = F$$

Most blood flow is laminar, meaning that the flow is considered to be smooth. Specifically, this behavior is observed with the speed of the blood in small arteries and capillaries. pH regulates the blood at 7.4, and the overall blood flow of an adult at rest is 5000 milliliters per minute (ml min). Flow rate in a cylindrical tube such as a blood vessel is contingent on the pressure difference, the dimensions of the tube and the viscosity of the fluid. **Viscosity** is an internal friction between neighboring layers of fluid as the layers migrate past one another. It is due to electrical, cohesive forces between molecules. Thus, the amount of thickness is addressed and a difference in pressure between the ends of a level tube is required for steady flow. When a fluid is viscous, energy is required and friction between molecules occurs. A liquid is typically incompressible, whereas its volume will not change; however, gases are amongst the fluids that are compressible. For a known fluid, the required force is proportional to the area of the fluid and the speed, and inversely proportional to the distance that separates the walls/tubes. This is shown in the formula below:

$$F = \eta A \frac{v}{l}$$

where F is the amount of force in Newtons, η is the coefficient of viscosity and is dependent on temperature and the type of fluid, A is the area, v is the velocity and l is the distance or amount of separation between the two layers. The more viscous a fluid, the more force is required for it to move. Whole blood at 37 °C has a coefficient of viscosity of ~ 4×10^{-3} Pa·s, whereas blood plasma's coefficient of viscosity is 1.5×10^{-3} Pa·s. Thus, force and viscosity are directly proportional to each other.

Ohm's law ($V = IR$) was discussed in chapter 4 and can be correlated with respect to blood flow. Just as a higher current implies less resistance, more blood flow suggests a lower resistance. Blood flow is determined by a pressure difference and an inhibition of flow (or resistance) through the vessel. Since viscosity functions similarly to friction, a variance in pressure is needed for continuous blood flow. The pressure difference forces blood to be pushed through the blood vessels. Blood flow can be quantified based on Poiseuille's law:

$$Q = \frac{\pi R^4 (P_1 - P_2)}{8 \eta L}$$

where Q is the rate of blood flow, η is the coefficient of viscosity, R is the radius of the blood vessel, L is the length of the vessel, and $P_1 - P_2$ is the pressure difference. This formula demonstrates that the diameter of the blood vessel plays a tremendous role in regulating the rate of blood flow through a vessel.

A person's posture can also impact the cardiac function. In a reclined position, the heart and the brain are leveled and there is less restriction on the heart's ability to pump the blood. A perfect example of this would be the adjustable beds that allow a person to elevate both their head and feet to their desired position. Yet, the outcome varies when a person is stationary, sitting or standing. In a static position, there is an increased 'load' and blood flow is impeded. If a person was required to be on 'bed rest' due to some condition, they may develop bed sores or simply have very poor circulation due to the lack of movement. When a person is sitting or standing, centrifugal force can press down on the heart and obstruct blood flow from the heart to the brain. In general, people with high blood pressure are able to tolerate greater amounts of centrifugal force because their heart works at a higher rate than someone that has a lower blood pressure. Also, shorter people tend to have a greater tolerance for gravitational force because the distance between the heart and brain is less compared to a person who is taller. When there is a delay in blood delivery to the brain, a person may blackout or simply become unconscious. A loss of consciousness typically leads to fainting and the person ends up in a prone position. This allows a more convenient way for the heart to deliver blood to the brain.

6.2 Membrane potential

Electric potentials are present across the membranes of most cells. The membrane functions as an insulator. Discrepancies between the electrical potential of the interior and exterior of a biological cell is the membrane potential. The exterior membrane potential ranges from −40 to −80 millivolts (mV). The membrane potential has two primary functions: (1) to authorize a cell to operate as a battery, providing power for a variety of molecular processes in the membrane; (2) to transport signals in electrically excited cells. Variations in ion concentrations across a selectively permeable membrane can produce a membrane potential.

Chloride, potassium and sodium ions are the key ions concerning the development of membrane potentials. Potassium regulates heartbeat and reduces blood pressure. It is essential in conducting electrical impulses to tell muscles to move. Low amounts of calcium can result in irregular heartbeat. Heart cell rhythm is dependent on the opening and closing of ion channels. When cardiac cells beat, sodium channels open, allowing an expedient flow of sodium ions into the cells as quickly as 2 milliseconds. Following this depolarization of the membrane, a release of potassium ions to the outside of the cell results within 0.5 s—the cell membrane repolarizes.

The heart has the unique ability to 'spontaneously depolarize', lacking the requirement of external stimuli to create a positive increase in voltage across the cell's membrane. This instinctive depolarization is called autorhythmicity. Autorhythmicity happens because the heart's membrane has a lower permeability to potassium; yet, passive transfer of calcium ions takes place. Cardiac cells are distinctive in the fact that they are the only cells within the body that are able to contract without being stimulated by the nervous system. When the cell is not being stimulated, a resting membrane potential exists. The standard resting membrane potential in the ventricular myocardium ranges between −85 and 95 millivolts (mV).

6.3 Electrocardiogram and echocardiogram

The electrocardiogram or ECG uses electrodes to study the electrical activity of the heart, while the echocardiogram measures the sound waves of the heart. ECG monitors the position and size of the chambers in the heart, as well as the beats of the heart and the presence of any damage. If a person takes any type of medication for their heart or has an implanted device, the ECG can detect their existence. If you have ever made a reference and heard of 'lub-dub', this is the sound that the heart makes as it contracts and relaxes. Since the ECG assesses the heart rate and rhythm at the time of measurements, it is possible that this technique will miss an intermittent cardiac abnormality. The tracing for an ECG includes a P, QRS and T wave, as shown in figure 6.2. These tracings can provide information about the heart in regards to the heart rate and rhythm, how electrical impulses spread across the heart, thickening of the heart muscle or whether there may be the presence of coronary artery disease. The P wave exemplifies when the atria depolarize, lasting nearly 80 milliseconds (ms). The atria contract and pump blood into the ventricles. Following, the ventricles depolarize, generating the QRS wave within 80–100 ms. The amplitude of the QRS wave is significantly larger than the P wave because the

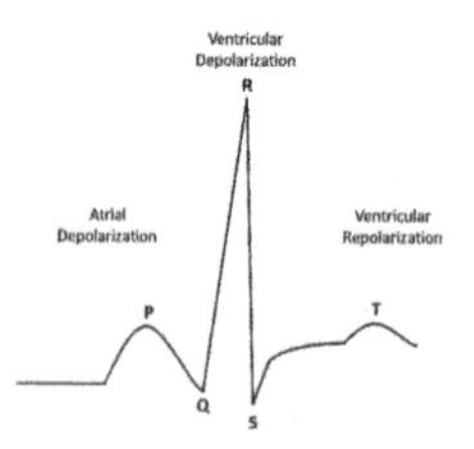

Figure 6.2. Electrocardiogram (ECG). During a normal ECG, the P wave is when the atria depolarize, the QRS complex is when the ventricles depolarize, and the T wave is the repolarization of the ventricles.

ventricles have a larger muscle mass than the atria. The T wave represents when the ventricles recover. While the ECG is efficient at measuring the heart's electrical behavior, it is not an accurate technique to gauge the pumping ability of the heart.

An echocardiogram (echo) is a non-invasive technique that utilizes sound waves to generate detailed pictures of the heart. The echo provides an indication of the pumping ability of the heart, and is most commonly used to diagnose and manage heart diseases. It measures the size and shape of the heart, as well as the internal structure of the heart and the velocity of the blood. The echocardiogram can be used to identify the location of tissue damage. The stress echocardiogram can aid in identifying if a person has heart-related chest pains. Both normal and abnormal blood flow through the heart can be measured with the echo, utilizing pulsed or continuous wave Doppler ultrasound. A major advantage of the echo is that there are no known risk factors or side effects associated with this measurement.

6.4 Cardiac abnormalities

In the event that the SA node misfires, an extra heartbeat or palpitation may arise. An arrhythmia is simply an abnormal rhythm of the heart, affecting its electrical activity. A palpitation may happen along with an arrhythmia; yet, palpitations can occur without being accompanied by an arrhythmia. For instance, if you feel your heart 'pounding', it may simply mean that you consumed too much caffeine or that you are stressed; this will accelerate your heartbeat. When the resting heartbeat is less than 60 bpm, it is called bradycardia; tachycardia is a heartbeat faster than 100 bpm.

If current travels across the chest from hand-to-hand and passes through the heart, fibrillation can occur. Fibrillation is when the heart muscles independently move in a chaotic fashion, versus in a normal, organized manner. This of course will impact the heart's ability to properly pump blood, resulting in cardiac arrest and brain damage. Both AC and DC can initiate fibrillation; it takes 300–500 mA for DC, and 30 mA (with a frequency of 60 hertz) for AC. A much lower current of 1 mA can lead to fibrillation if the current travels directly to the heart. Fibrillation can be deadly if not treated with defibrillation, because circulation is disrupted. A heart defibrillator is a charged capacitor with a high voltage of a few thousand volts. In the case of a heart attack, irregular heartbeats occurs and a defibrillator may be useful. A momentary jolt of charge from the defibrillator will 'shock' the heart, allowing the cells to recharge and instigate regular beating. The defibrillator has paddles that distribute current over the chest and discharge voltage very quickly through the heart.

There are various types of cardiac arrhythmias that affect the atria and ventricles including atrial fibrillation and ventricular fibrillation. Atrial fibrillation (AF) is the most common of all arrhythmias, where a heartbeat can range between 350 and 500 bpm. AF may result from high blood pressure or excessive alcohol consumption, and typically accompanies another existing medical abnormality. While the electrical impulses associated with AF transpire at a very rapid pace, heartbeats are uncommon. The AV node slows down the conduction velocity and protects the ventricles. Neuronal-type sodium channels are thought to mediate sodium-calcium crosstalk within nanodomains that contain calcium release, and ultimately instigate AF. Conversely, ventricular fibrillation (VF) can be fatal. With this arrhythmia, erratic electrical impulses occur and the ventricles 'quiver' instead of pumping blood. The blood pressure significantly declines, there is no blood supply to the vital organs and the person collapses within seconds. A heart attack is the most common source of VF. However, VF may happen if the heart muscle has insufficient amounts of oxygen or if there are disturbances in electrolytes. VF is considered to be a form of cardiac arrest. In the event that a person does not recover within 2–3 min of this arrhythmia, death is highly probable.

Heart failure is when the heart loses the ability to fill and empty—there is insufficient blood flow. Defective pumping may cause fluid retention in the lungs and shortness of breath—known as congestive heart failure. The primary source of heart failure is coronary artery disease (CAD). CAD is the narrowing of the arteries. The resting coronary blood flow for a person is approximately 225 ml min^{-1}. During CAD, the rate of blood flow is depleted. CAD is also a major cause of myocardial infarction or heart attacks. This is due to the fact that the heart continuously has the desire to contract. Unfortunately, the heart does not get a full contraction, adequate volumes of blood do not reach the body and oxygen deprivation occurs. During a heart attack, irreversible death of the heart muscle develops if blood flow is not restored within 20–40 min. The muscle continuously dies for hours at a time and is replaced by scar tissue.

Further reading

Ferrier G R and Howlett S E 2001 Cardiac excitation-contraction coupling: role of membrane potential in regulation of contraction *Am. J. Physiol.-Heart and Circ. Physiol.* **280** H1928–44

Giancoli D C 2016 *Physics: Principles with Applications* (Boston, MA: Pearson)

Kaufman W and Johnston F D 1943 The electrical conductivity of the tissues near the heart and its bearing on the distribution of the cardiac action currents *Am. Heart J.* **26** 42

Libby P and Theroux P 2005 Pathophysiology of coronary artery disease *Circulation* **111** 3481–8

McCall R P 2010 *Physics of the Human Body* (Baltimore, MD: Johns Hopkins University Press)

Munger M A *et al* 2020 Tetrodotoxin-sensitive neuronal-type Na+ channels: a novel and druggable target for prevention of atrial fibrillation *J. Am. Heart Assoc.* **9** e015119

Stinstra J G, Hopenfeld B and MacLeod R S 2005 On the passive cardiac conductivity *Ann. Biomed. Eng.* **33** 1743–51

IOP Publishing

Biophysics of the Senses (Second Edition)

Tennille D Presley

Chapter 7

The human mind

Similar to the heart, the brain is a component vital to the body. It facilitates networks that ultimately drive processes throughout the body necessary to survive and function—the brain controls our senses and many biological processes. As the saying goes, 'A mind is a terrible thing to waste'. This simplistic yet profound statement speaks volumes because the mind is so inimitable and powerful—ultimately, we only get one and have to feed it, exercise it, and take care of it to the greatest extent possible. In this chapter, a neuron is presented; it is responsible for conveying information from one portion of the brain to the other. Nerve impulses then orchestrate communication between nerve cells. To aid in a better understanding of the brain function, quantum mechanics is discussed. The 'thinking' power of the brain (and a lack there of) lead into the essential role of networks and how the brain provides regulatory control.

7.1 What is a neuron?

The human mind is often related to the organ that makes up nearly 2% of the body's weight, the brain; however, the human mind is influenced by other organs in the body in addition to the brain. Have you ever wondered why you desire the food you smell or if you see something on television, you want to go and buy it? When the mind is impassable, it is similar to a closed circuit where the flow continues and there is no entryway without completely disrupting the normal flow. The mind is best represented by Newton's third Law of Motion which states '*For every action, there is an equal-opposite reaction*'. When you exercise the brain, the brain will react and become stronger.

A neuron is a large nerve cell that receives and transfers information from one part of the brain to the other. It entails a cell body and two types of extensions. These extensions are axons and dendrites; the axons normally transmit signals away from the cell body, whereas the dendrites transport signals toward the cell body. Axons have a smooth surface and there is only one axon per cell. These long nerve

doi:10.1088/978-0-7503-3283-5ch7

processes end at synapses. On the other hand, dendrites are smaller, more abundant than axons, and have many synapses to receive messages from surrounding neurons. Dendrites typically have more branches and have a rough surface. Axons and dendrites are bundled to comprise nerves. Nerves transport signals between the brain, spinal cord and other parts of the body through nerve impulses. These nerves function for thinking, muscle movement, and to relay information to and from the senses. The cell body entails the cell membrane, the nucleus of the cell and control the rate at which the electric signals travel through the neurons.

The brain houses nearly 100 billion neurons, and the central nervous system (CNS) contains roughly the same amount. There are different types of neurons. Neurons are considered as motor neurons or sensory neurons. Motor neurons are responsible for conveying information from the CNS to glands, muscles and organs throughout the body. For instance if you are ready to leave your home, your motor neurons tell you to move your legs, to walk to the door and turn the doorknob to exit. On the other hand, sensory neurons receive information from external stimuli or internal organs and transfer that information to the CNS. They are motivated by temperature, pain, light and sound, where these interactions are communicated to the brain. If you are outdoors and it is 20 °F, your sensory neurons will send the message to your CNS to let you know that you feel cold. Interneurons are the communicators between the motor and sensory neurons, and represent more than 99% of the neurons in the body. They 'bridge' the neuronal communication between the brain and the spinal cord. Changes in electrical properties trigger the transport of information along the axons and dendrites. When the terminal end of an axon receives an electrical signal, neurotransmitters are released to connect with other neurons and cells.

When functioning properly, ions migrate across the cell membrane of a neuron. Through this process, energy is transferred and a membrane potential is generated. There are instances where 4000 new neurons are created every second. These neurons eventually form networks, and become involved in the communication between the brain and the rest of the body. When a stimulus is applied, the ions move perpendicularly to the axon. This causes a change in the electric potential instigating the movement of ions through the membrane. This differs from the electrical current in wire, as its path is along the same direction of the wire. Recent studies suggest that the aging brain has the capacity to regenerate new brain cells, similar to younger individuals—this is believed to occur in the hippocampus, which is responsible for memory, cognition and emotion. Both the neurons and the blood vessels partner to facilitate this process.

7.2 Nerve impulses

As previously defined, an impulse represents the product of a force and time. When thinking of a nerve, an electrical signal travels along an axon known as a **nerve impulse**. A nerve impulse is a wave of electrical activity that travels between nerve cells. Chemical processes create electric energy in the nerve cells, and energy fields result around the body. Billions of nerve impulses of the same size migrate throughout the brain and the CNS. The frequency of the impulses regulate the intensity of each nerve signal. An electrical difference between the surrounding space and the inside of the axon exists, similar to a battery. Once the nerve is activated, a change in voltage occurs across the wall of the axon and ions flow in and out of the neuron. The speed for a nerve impulse can vary depending on the type of neuron. The fastest nerve impulses can travel as rapidly as 250 miles per hour. There are nearly a billion nerve impulses in the body that regularly produce magnetic fields. An axon must be properly insulated for an impulse to migrate swiftly. Myelin is a fatty substance that provides insulation to the axon. While this utilizes a lot of energy, there are instances where a neuron may need to transfer information immediately. For example, if you are placing a dish in a hot oven and accidentally touch the oven rack, it is critical that your brain quickly receives the message that you are burning your hand and to remove your hand instantly. Alternatively, myelin may begin to degrade such as with multiple sclerosis. In multiple sclerosis, the nerves are unable to effectively carry electrical signals between the brain and the body.

There is a higher concentration of potassium ions inside the cells and sodium ions outside of the cell. Because of this, there is a tendency for the ions to diffuse from a higher concentration to a lower concentration. The potassium ions will tend to move outside of the cell and the inside of the cell becomes more electrically negative. The resting membrane potential is −70 mV. The presence of the membrane potential causes an electric force to act on the ions. Nerve signals rapidly spread along the nerve membrane by action potentials. The gray matter of the brain is responsible for collecting and conducting nerve impulses.

Electroencephalography (EEG) is used to record the ionic current flowing through the neurons. It measures the voltage changes in the electrical activity along the scalp, and the electrodes detect the electric field generated by moving charges in the connections between the neurons (synapses) in the brain. However, it is not useful for diagnosis of a headache. The brain does not have pain receptors; however, there are regions of the head and neck that do. Thus it is called a 'headache' and not an aching brain.

7.3 Quantum mechanics

We have explored the aspect of mechanics, but mainly focused on classical mechanics in chapter 2 and touched the surface of quantum mechanics in chapter 4; however, this aspect of physics plays a major role in the human mind. Do you ever stop to think about the things that exist, but are so small that you cannot see them with the naked eye? Quantum mechanics is an aspect of physics that studies the 'very small', addressing the world of atoms and light. Quantum mechanics is also important for understanding higher brain functions such as the generation of voluntary movements, and it is believed that a correlation exists between consciousness and quantum theory.

Our sense of sight is a series of events between the eye, the brain and the outside world. In general, a person rapidly moves their eyes in a coordinated fashion that activates billions of neurons in the brain. The retina in the eye has a sensitivity to small numbers of photons, which are particles of light. The photon can then interact with the quantum state of the retinal cell. Light energy is detected by the eye, and information about intensity, color and shape is transmitted to the brain. The light is converted into electrical signals and the brain interprets those signals into images. This impacts perception, which is the process by which the brain interprets and organizes sensory information. How do you feel when you listen to your favorite song or you hear something that is inspirational? Your senses allow you to pick up this sound and your sensory neurons utilize your networks to cause you to feel 'positive energy'.

Consciousness is the state of being self-aware and having sensory experiences. It has been speculated that quantum fluctuations inside of microtubules create consciousness. Microtubules contain units of tubulin that is made up of regions where electrons spin closely to each other (though not touching), and ultimately affect each other—quantum entanglement. When conscious awareness goes below the absolute threshold, subliminal stimulation occurs. An example of this is subliminal advertising. Subliminal advertising is a hidden agenda that triggers electrical signals in your brain creating a desire and want for something that you did not ordinarily want. There is now evidence that invisible, subliminal images subconsciously attract the brain's attention. The brain can be affected by images that you are not even conscious of seeing.

7.4 Cognition and neurological disorders

Cognition is simply the thinking processes of the brain related to attention, memory, evaluation, reasoning and problem solving. Cognitive dysfunctions may cause a person to become forgetful or lose some of the normal mobility that a person naturally has. These dysfunctions may be induced by chemical activity, electrical activity or other forms of external factors. For instance, a lack of magnesium can lead to poor memory function, and a shortage of phosphate may result in mental confusion and problems with speech. Similarly, excessive levels of calcium can contribute to depression and confusion. Sleep is a way to restore the body's energy and restore memory.

Other aspects of neurological abnormalities also exist. Think of a concussion or a tremor. A concussion is the most common traumatic brain injury and occurs when there is a certain amount of force applied to the skull; it leads to modifications in the normal function of the brain. A tremor is an involuntary movement of muscle contraction and relaxation. It can imply stress or overconsumption of caffeine; in some instances, it is an indicator of a neurological disorder such as traumatic brain injury, stroke or multiple sclerosis. Multiple sclerosis is an inflammatory process that is characterized by the destruction of the myelin sheaths (the protective, insulating outer coverings of the nerves) within the CNS. Symptoms of this disorder include muscle stiffness, double vision, mental confusion and loss of memory (in severe cases).

There is also the comparison of a coma versus being 'brain dead'. During a coma, a person is unconscious and non-responsive; yet, the individual still has some brain activity and is considered 'alive'. In the presence of a coma, there may be brain stem responses, non-purposeful motor responses and/or spontaneous breathing. Comas can be temporary or permanent, having three potential outcomes: (i) recovery of consciousness; (ii) chronically depressed consciousness (i.e. a vegetative state); (iii) brain death. When a person is 'brain dead', the individual is considered to no longer be alive and there is completely no brain function. This is an irreversible process that leads to brain swelling, depleted blood flow to the brain, and tissue death. When the brain dies, all of the inner workings in the body cease—breathing nullifies and the heart no longer beats.

As the brain ages, it is more prone to diseases such as dementia or Alzheimer's. Dementia is a progressive decrease in a person's intellectual ability, where the brain's ability to think, reason and remember is significantly affected. Damage to the brain typically occurs for some time before there are any visible signs of this disease. Dementia is characterized by impaired judgement, memory loss and drastic changes in personality. Alzheimer's disease is the most common form of dementia and is when the brain ages prematurely. A common finding of this disease is that neurons are lost where cognitive information and memory processing occurs. This disorder can lead to depression, short-term memory loss, a lack of effective communication, and urinary incontinence.

Electrical current can cause neuropathy. Electrical injury can result in neurocognitive dysfunction, potentially altering a person's memory, concentration, attention span and speed of mental processing. If the current travels through the head, loss of consciousness typically happens. Victims of electrical shock have exhibited functional differences in neural activation during spatial working memory and implicit learning ocular tasks.

7.5 Networks and regulatory control

In our current era, we are amidst many venues of social media from Facebook to Twitter to Instagram, just to name a few. These popular mechanisms have developed a way for people to connect and reconnect. As a person connects with another, it is possible to see other people that each of them are connected with. Networks are established, which is very similar to how the networks in the body exist. The web is a perfect analogy of the body because it has a plethora of networks that exist to provide function and communication. The brain is the central conduit for regulating all of the networks of the body. It is through these networks that we move, breath, smell, feel, see and hear. Without the brain, none of the networks in the body will work.

Within the suprachiasmatic nucleus (SCN) of the brain, a master circadian clock exists that aids to direct the body to sleep at night and to be awake during the day. Containing nearly 20 000 nerve cells, the SCN contributes to regulating temperature and hormone levels in the body. Circadian rhythms are naturally produced, but they are affected by environmental factors such as light and darkness from the eyes to the SCN. This internal clock helps the body to 'harmonize' with a 24 h day. Additional time cues such as meal and exercise schedules can affect the timing of the clock.

As a person ages, circadian rhythms may change. Disorders in the circadian rhythm can result in insomnia, depression, bipolar disorder and impaired work performance. Circadian rhythms can be impacted by melatonin, which is created by the pineal gland in the brain. A benefit to the circadian clock is that the timing of a medical treatment in coordination with the body's clock can potentially decrease drug toxicity while enhancing efficacy.

Electroconvulsive therapy, or ECT, is a technique that involves an intentional seizure being triggered by electric currents that are transmitted through the brain. ECT triggers modifications in brain chemistry, and is often successful when other treatments are not. It is used to treat severe depression, severe mania or even aggression in people with dementia. Although ECT is typically safe, some of the side effects of this treatment are confusion, memory loss, nausea, headaches and vomiting. Yet, these detriments are usually short-lived.

The brain and the nervous system generate magnetic fields. Has another person ever made you feel more energized, excited or completely drained? This is in part due to the fact that a person's electromagnetic field can influence their surroundings. In general, your senses inform your brain about variations in your surroundings. Thus, the feeling of your shoes on your feet, the sound of your fish aquarium, the sound of the computer or refrigerator humming, may often go unnoticed. Alternatively, your brain allows you to focus on things that really capture your attention and interest such as gazing into the eyes of your significant other, watching your favorite television show or being engaged in class. The brain 'illuminates' the things that are of great importance. Even before a person pays close attention to something of interest, the irrelevant, background 'noise' is filtered out. Most of the sensory neurons are unipolar, where the receptor region is housed in the peripheral nervous system to control the sense of touch, pain or temperature. Bipolar neurons are less common, but are localized in the eye, ear and nose.

As a whole, the body is full of networks that power the human mind, where the brain is the 'master regulator'. These networks work together to regulate control throughout the body, utilizing each of our senses.

Further reading

Becker R O 1990 *Cross Currents: The Perils of Electropollution, the Promise of Electromedicine* (Los Angeles, CA: Jeremy P Tarcher)

Becker R O and Selden G 1985 *The Body Electric: Electromagnetism and the Foundation of Life* (New York: Quill)

Bagasra O, Michaels F H, Zheng Y M, Bobroski L E, Spitsin S V, Fu Z F, Tawadros R and Koprowski H 1995 Activation of the inducible form of nitric oxide synthase in the brains of patients with multiple sclerosis *Proc. Natl. Acad. Sci. USA* **92** 12041–5

Hall J E and Hall M E 2020 *Guyton and Hall Textbook of Medical Physiology e-Book* (Amsterdam: Elsevier Health Sciences)

Herman I 2007 *Physics of the Human Body* (Berlin/Heidelberg: Springer)

Koch C and Hepp K 2007 *The Relation Between Quantum Mechanics and Higher Brain Functions: Lessons from Quantum Computation and Neurobiology* (Pasadena, CA: California Institute of Technology)

McCall R P 2010 *Physics of the Human Body* (Baltimore, MD: Johns Hopkins University Press)

Tartt A N, Fulmore C A, Liu Y, Rosoklija G B, Dwork A J, Arango V, Hen R, Mann J J and Boldrini M 2018 Considerations for assessing the extent of hippocampal neurogenesis in the adult and aging human brain *Cell Stem Cell* **23** 782–3

http://livescience.com/37807-brain-is-not-quantum-computer.html

IOP Publishing

Biophysics of the Senses (Second Edition)

Tennille D Presley

Chapter 8

Physics of nutrition, exercise and disease

As described in chapter 7, the brain has a major impact on the body and is the navigator of the senses. Taste is no different as it is involves electrical signals that are sent to the brain. It is no secret that the body is fueled by food and physical activity; combining these aspects with thermodynamics can further enhance the benefits of each effect. In this chapter, the importance of maintaining a healthy diet and regular exercise is discussed. Contributing factors including the regulation of free radicals, balancing calories, and the conservation of energy (i.e. simple machines) are explained in correlation to nutrition and exercise. Both hot yoga and heat acclimation are techniques that combine exercise with moderate exposure to elevated temperatures; each of their attributes are introduced. When the body malfunctions, diseases may occur. The biophysics associated with diseases such as sickle cell disease and diabetes are discussed.

8.1 Physics of a healthy diet—sense of taste

Taste buds are 'chemoreceptors' located on the upper portion of the tongue in the mouth that have the ability to detect taste. This sense involves chemical signals in foods and electrical signals within the body. The electrical signals migrate to the brain via the nervous system, informing the brain of the sense of taste. Taste buds make direct contact with the chemicals in order for us to taste; however, indirect contact with chemicals occurs for us to smell. The taste buds influence the human mind. An adult has between 3000 and 10 000 taste buds, and this value degrades with age. There are five distinct tastes: (1) salty, (2) bitter, (3) sweet, (4) sour, (5) spicy. Each taste bud typically responds to only one of the 'distinct tastes' when there is a low concentration of a substance; however at high concentrations, most taste buds will be stimulated by practically all of the tastes. Foods that are high in calories are usually salty, sweet or savory. Acidic foods (pH < 7) are typically sour or bitter. The average person uses roughly 2000 calories each day in food. The accumulation of a radical can often shift a substance from sweet to bitter.

doi:10.1088/978-0-7503-3283-5ch8

Acidic foods tend to taste sour, whereas basic foods have a bitter taste. An acidic fluid that exhibits benefits of health is apple cider vinegar. It attenuates blood pressure, augments weight loss, and combats disease.

Consider the coronavirus (COVID-19) for example—a common symptom of the virus is a loss of taste and smell. Amongst many interviews of individuals that had COVID-19, there were numerous comments where people learned that eating is often based on taste and not so much hunger. Several people experienced weight-loss while having COVID, and this was due in part to the lack of the desire to eat, mainly due to the inability to taste and enjoy the food.

One of the major facets of 'dieting' is to count the calories. In chapter 3, the calorie was introduced and we identified that 1 Calorie = 1 kilocalorie. As we read food labels and identify the amount of 'calories', the reality is that these are actually 'kilocalories'. For example, the popular '100 calorie' snacks are really 100 000 Calorie snacks. When we eat, the body turns the food into energy. The major sources of energy in foods are carbohydrates, proteins and fats. One gram of carbohydrates is equivalent to 4 calories, 1 gram of protein is equal to 4 calories and 1 gram of fat contains 9 calories. The amount of energy that is gained from a food can be quantified by multiplying each one of these constants by the amount (in grams) of fats, carbohydrates and proteins that are contained in the food that will be consumed. Having a grasp of these calculations aids in properly reading and understanding food labels. Starvation or a low-restrictive diet can cause a 20%–30% decline in a person's basal metabolic rate. In chapter 4, the components of an atom were discussed. In a situation where there are the same number of protons but a different number of neutrons, an isotope exists. Stable isotopes can be used for examining the flow of nutrients through the body. The standard isotopes are calcium, iron, magnesium and zinc. When your body craves certain foods, it is often implying that the body is deficient of what it is craving.

Since water lacks calories, it is an easy way to manage daily caloric intake and maintain a healthy diet. It should be the main source of hydration and a major contributor to managing daily weight gain. As we age, the body composition of water becomes depleted. For instance, a newborn's body is comprised of nearly 75% of water; however, the body of an elderly person is made up of 50% water. If the body contains more fat, it contains less water. Yet, having more muscle means more water. In addition, the fundamental organs of the body have different amounts of water. The brain, heart, liver and kidneys are mainly water, comprising between 65%–85%, whereas the bones are made up of approximately 30% of water.

We produce more free radicals when consuming fatty foods. To reverse this effect, dietary nitrates have been shown to be advantageous. Eighty percent of dietary nitrates arrive from vegetable consumption. These vegetables include spinach, celery, carrots, and beets. Dietary nitrates account for half of the steady state concentration of nitric oxide, and aid in reducing blood pressure, improving intestinal health and exercise performance. Specifically, beet juice contains high levels of nitrates and has been shown to enhance the exercise capacity of younger and older adults, as well as individuals with chronic obstructive pulmonary disease (COPD). Nuts have similar effects as dietary nitrates as they are a great source of antioxidants, specifically vitamin E; they protect against cognitive decline. Likewise, blueberries and avocados promote healthy blood flow and inhibit oxidative stress.

The foods that a person consumes can be compared to a running capacitor versus a starting capacitor. A starting capacitor will help with starting a motor and then it is no longer a part of the circuit. However, a running capacitor has to be a part of the circuit at all times. In the body, the heart and the brain are the running capacitors because both play vital roles in regulating control in the body. However, an energy drink or foods can act as starting capacitors to get the body going, and then the body can function. It is possible to see a decline in capacitance from weight loss. The body can survive without food for nearly three weeks on average, but water is necessary at least every three to four days.

8.2 Exercise

It is no secret that regular physical activity is important for both mental and physical health. It is believed that high intensity interval training allows your metabolic rate to be active throughout the day, while cardio training is only within that particular instance. The concept of work 'conservation' is key with exercising. We know that work in defined as a product of work and displacement. The aspect of simple machines expands the concept of work by denoting that *the amount of work into a system is equivalent to the amount of work out of a system*. Thus,

$$\text{Work input} = \text{Work output}$$
$$(\text{Force})_{\text{in}}(\text{distance})_{\text{in}} = (\text{Force})_{\text{out}}(\text{distance})_{\text{out}}$$
$$W_{\text{in}} = W_{\text{out}}$$

Example 8.1. While competing in an obstacle course, Helen pulls a rope downward 3.25 m with a force of 334 N, to lift herself. If she weighs 132 lbs, how far does Helen travel upward?

Solution: Let's first consider what we are given

$$d_{\text{in}} = 3.25\text{ m}$$
$$F_{\text{in}} = 334\text{ N}$$
$$w = F_{\text{out}} = 132\text{ lbs} = 587.2\text{ N}$$
$$d_{\text{out}} = ?$$

Recognizing that this is a simple machine problem, consider the following equation

$$(\text{Force})_{\text{in}}(\text{distance})_{\text{in}} = (\text{Force})_{\text{out}}(\text{distance})_{\text{out}}$$
$$(334\text{ N})(3.25\text{ m}) = (587.2\text{ N})d_{\text{out}}$$
$$1085.5\text{ N}\cdot\text{m} = (587.2\text{ N})d_{\text{out}}$$
$$\mathbf{1.\ 85\ m = d_{out}}$$

In chapter 2, levers were discussed in regards to how the muscles and joints function as a lever. A lever is a simple machine—a device for modifying the direction of force. Some examples include levers, pulleys, and wheel and axles. The underlying principle to simple machines is the conservation of energy. Example 8.1 provides an example of a simple machine. Depending on the type of exercise you carry out determines which type of simple machine is relevant. For instance, the lever is a perfect example of how resistance and weight training is performed.

When lifting heavy weights, a person should lift the weight close to their body to lessen the torque generated around the lower spine. Resistance training provides continuous tension on the muscles, working them under a constant load. Each time a person lifts weights, they are lifting with an upward force that opposes the force of gravity. The major advantage that resistance bands have compared to free weights is that the bands' resistance is not dependent on gravity. Because of this, it is possible to have more functional movement patterns and a better range of motion that mimic daily activities.

Exercise boosts energy, promotes better sleep, weight loss, and improves mood. The pedometer is a useful tool because it allows an individual to count their number of steps on a day-to-day basis. As highlighted in table 8.1, there are a number of daily activities that can burn calories. For example, going to the grocery store can burn up to 200 calories depending on the amount of time you spend in the store, and how much you 'squat' to pick up items. Cleaning the house can also be beneficial to your health, especially if you decide to mop the floor which can burn up to 400 calories.

The way that a person floats in water is dependent upon the force of gravity and the buoyant force. Buoyancy is the apparent loss of weight of an object submerged in a fluid. The buoyancy of water decreases a person's body weight by nearly 90%. When a person floats, it means that their buoyant force and gravitational force are equal, yet opposite to each other. Swimming is a great exercise that incorporates the concept of buoyancy, and is a full-body workout. It is very beneficial as it is low impact, provides a natural resistance and promotes cardiovascular health.

Table 8.1. List of daily activities and the amount of calories each burns.

Activity	Number of calories burned
Gardening	600 calories
Sleeping (for 8 h)	350 calories
Aerobic dancing (1 h)	300 calories
Mopping the Floor	400 calories
Mowing the Lawn	300 calories
Grocery shopping	200 calories
Ironing for 30 min	75 calories
30 min of kissing	36 calories

For example, a normal heart rate ranges between 60 and 100 bpm (as highlighted in chapter 6); however, this rate is significantly reduced to 17 bpm when swimming.

One of the latest trends is the hamster-wheel desk. This is the best motto of 'exercise while you work'. People are very consumed with work on a day-to-day basis; however, this new invention provides the opportunity to exercise while working. The advantage to this wheel versus a treadmill is that you can easily set your own pace while you work or to stand still. As the wheel rotates in a vertical circle, the centripetal force acts and the person walks at a slight angle with each step.

We learned in chapter 3 that heat is a measure of energy in transit. Over the last several decades, studies have shown how the appropriate exposure to heat, with the optimum temperature can provide beneficial effects to the body. Hyperthermia or heat shock is when the body is exposed to an elevated, non-physiologic temperature for a short period of time. The temperature of the surrounding environment is increased; however, the body's core temperature typically remains physiologic. Hyperthermia has been suggested as an alternative for individuals who are unable to adequately exercise. Other advantages of heat shock include increased thermotolerance, decreases in ROS, and improvements in the prevalence of both nitric oxide and heat shock proteins (hsps). Hsps are intracellular soluble proteins that act as chaperones to ensure that cells maintain the right shape and position. Named on the basis of their molecular weight, hsps are also cardioprotective. A key benefit of acute heat is that it overexpresses hsps. Some hsps (i.e. hsp90) are able to conjoin with variations of nitric oxide to reduce hyperglycemia by way of downregulating glucose transporter 1 (glut-1) and upregulating glucose-6-phosphate dehydrogenase (G6PD). Glut-1 supports glucose transport across the plasma membrane of mammalian cells, whereas G6PD protects red blood cells from damage and destruction. Heat acclimation is exercising in heat (~40 °C–43 °C) for a brief period of time. The advantages include a better regulation of blood pressure, increases in nitric oxide and an improved exercise performance. The time of duration for this process is critical and is most commonly observed in individuals that are considered to be elite, athletes. In each instance of both hyperthermia and heat acclimation, it is important that the body acquires a significant amount of fluids to avoid dehydration.

Hot yoga is when numerous yoga styles are used in the presence of heat (temperature ranging from 90 to 108 degrees Fahrenheit) to enhance a person's flexibility in the poses. Depending on the pose, specific parts of the body lengthen, while other parts of the body contract to maintain stability. Due to the heat, the poses are usually easier and detoxification happens. Yoga also improves brain function, lowers stress levels, decreases blood pressure and risks for heart disease. Furthermore, hot yoga amends arterial stiffness and improves quality of life, especially for overweight and obese individuals. Throughout this text, we have been discussing force. In the event that a force is *too great* or an object is overly stretched, a fracture will occur. When something is stretched beyond its limit, permanent deformation results.

When strenuous activity or overexertion occurs, normal body functions are compromised and inadequate levels of electrolytes are present. Energy drinks such as Gatorade have electrolytes that can be useful when this deficiency transpires. The electrolytes can improve an athlete's physical performance and endurance. Supplements of nitric oxide are also known to be beneficial in improving exercise performance. Distance runners and elite athletes typically do not handle gravitational forces as well as a person who is not as physically fit. This is because individuals who are fit tend to have a lower blood pressure and the blood vessels are often more flexible. Elastic arteries can establish less of a pressure difference, and store blood. Thus, less blood is forced to flow through the body.

Iron is of great importance for athletic performance. It is needed for oxygen transport, energy production and cell division. A deficiency in iron can cause anemia and poor oxygen delivery to the muscles that are being exercised. Alternatively, cell and tissue injury results when there is too much iron. When there is an excessive amount of iron, the hydroxyl radical is formed.

8.3 Impact of disease

Biophysics is heavily related to disease diagnosis and treatment. Mechanics, electrical activity, free radicals and energy all relate to many of the diseases that currently dominate and affect the world. For instance, the microcirculation in both sickle cell disease and diabetes is impaired. During sickle cell disease (SCD), the mechanics of normal blood flow is dysfunctional. Anemia happens when there is a decline in hemoglobin. Typically, a red blood cell is flexible and can readily pass through the blood vessels and the capillaries. However, with SCD, the red blood cells are rigid and less deformable, causing painful episodes, oxygen deprivation and organ damage. To study this, researchers have utilized microfluidics. If polymerization occurs at a slow rate, the red blood cells can remain flexible and continue through normal circulation. However, polymerization at an 'intermediate speed' can be fatal. There is the likelihood that the cells will become rigid and get stuck in the vessel and cause obstructions in flow. In general, fluidics is important for gas exchange from the atmosphere and occurs through the nose, so that the cilia can act like a mechanical filter for impurities.

Type II diabetes is a major cause of death that is characterized by excessive glucose (hyperglycemia), insulin resistance and vascular dysfunction. Each of these detrimental effects, create chaos amongst the senses of the body from blurred vision to hearing loss. Similar to sickle cell disease, poor microcirculation due to decreased red cell deformability exists. Disorder in circadian rhythms and nerve pain have also been associated with diabetes. This disease can be controlled by diet and exercise. It is also believed that the probability of acquiring diabetes increases with age. The free radical theory of aging suggests that antioxidants will impede the process of aging by preventing free radicals from oxidizing sensitive biologicals molecules or reducing the formation of free radicals.

Whether it is the mechanics of the networks throughout the body or complete chaos due to excessive free radicals, an adequate appreciation of biophysics is key to maintain the normal function of the senses of the body. As a whole, the fundamental components of physics aid in our understanding of overall, efficient health.

Further reading

Aprelev A, Stephenson W, Noh H M, Meier M and Ferrone F A 2012 The physical foundation of vasooclusion in sickle cell disease *Biophys. J.* **103** L38–40

Belfield L T, Jeffers A B, Davis A T, Kavanagh K and Presley T D 2019 Biophysical regulation of thermodynamic effects on type 2 diabetes *AIP Conf. Proc.* **2109** 070003

Garrett A T, Goosens N G, Rehrer N J, Patterson M J, Harrison J, Sammut I and Cotter J D 2014 Short-term heat acclimation is effective and may be enhanced rather than impaired by dehydration *Am. J. Hum. Biol.* **26** 311–20

Hall J E and Hall M E 2020 *Guyton and Hall Textbook of Medical Physiology e-Book* (Amsterdam: Elsevier Health Sciences)

Hunter S D, Dhindsa M S, Cunningham E, Tarumi T, Alkatan M, Nualnim N and Tanaka H 2016 Impact of hot yoga on arterial stiffness and quality of life in overweight/obese adults *J. Phys. Act. Health* **13** 1360–3

Miller G D, Marsh A P, Dove R W, Beavers D, Presley T, Helms C, Bechtold E, King S B and Kim-Shapiro D 2012 Plasma nitrate and nitrite are increased by a high-nitrate supplement but not by high-nitrate foods in older adults *Nutr. Res.* **32** 160–8

Presley T D *et al* 2011 Acute effect of a high nitrate diet on brain perfusion in older adults *Nitric Oxide* **24** 34–42

Presley T, Vedam K, Druhan L J and Ilangovan G 2010 Hyperthermia-induced Hsp90· eNOS preserves mitochondrial respiration in hyperglycemic endothelial cells by down-regulating Glut-1 and up-regulating G6PD activity *J. Biol. Chem.* **285** 38194–203

Tytell M and Hooper P L 2001 Heat shock proteins: new keys to the development of cytoprotective therapies *Emerg. Therap. Targets* **5** 267–87

Wang Z, Deurenberg P, Wang W, Pietrobelli A, Baumgartner R N and Heymsfield S B 1999 Hydration of fat-free body mass: review and critique of a classic body-composition constant *Am. J. Clin. Nutr.* **69** 833–41

Weyer S, Ulbrich M and Leonhardt S 2013 A model-based approach for analysis of intracellular resistance variations due to body posture on bioimpedance measurements *J. Phys.: Conf. Ser.* **434** 012003

IOP Publishing

Biophysics of the Senses (Second Edition)

Tennille D Presley

Chapter 9

Music's influence on physics and the body

Music is a dynamic and universal language that is cross-cultural, and can impact a person's being. The way in which music is received is through the sense of hearing. As music enters the ear, you begin to feel it from the inside out. It may touch you in a certain way where you may feel emotional or it may cause your body to want to move and dance. Either way, music has a powerful impact on the human body. The process for hearing involves the outer ear, middle ear and the inner ear. When sound arrives at the outer ear, it acts like a 'funnel' and directs waves to the middle ear, causing vibrations. Once the vibrations arrive at the middle ear, they cause the bones to mechanically move—the vibrations are then sent to the inner ear. The inner ear picks up the mechanical vibrations and sends them to the brain as electrical impulses. This process is key in helping to understand the process for how music can affect the body. Music therapy is even a common facet that has been explored over the years to aid in managing various diseases. This chapter focuses on the evolution of the synergy between science and music, as well as how this crossover affects the human body in normal and diseased states.

9.1 History of sound and the science of music

Considered as the 'Father of Mathematics and Music', Pythagoras introduced the correlation between music and mathematics. Musical notes can be translated into mathematical equations. Geometry represents the mathematics, created at the atomic level of electrons as standing waves oscillate at assorted frequencies. Pythagoras felt that the Universe hummed with its own harmony, meaning that the Sun, Moon and planets emit their own distinctive sounds because of their orbital revolution. He taught that healing could come from sound and harmonic frequencies, and was the first to view music as a form of natural medicine. Pythagoras realized that strings with the same tension but varying lengths produced harmonious notes. Thus, the string theory was born: '*A string exactly half the length of another will play a pitch that is exactly an octave higher when struck or plucked. Split a string*

doi:10.1088/978-0-7503-3283-5ch9

into thirds and the pitch is raised an octave and a fifth'. If the string were split into fourths, an even higher pitch would be developed and so forth. The ratios are represented as follows: 2:1 is an octave; 3:2 is a fifth; 4:3 is a fourth.

Similar to Pythagoras, Johannes Kepler believed that each planet emits a tone with varying pitch as its respective distance from the Sun changes—this resulted in the production of 'music of the spheres'. Because the electron is the most spherical object in the Universe, Pythagoras believed that '*There is geometry in the humming of the strings, and there is music in the spacing of the spheres*'. Also known as *harmony of the spheres,* this phenomenon accounts for proportions in the movements of celestial bodies (i.e. the Sun, Moon, and planets) as a form of music. Planets move in elliptical paths with varying speeds, and those evolving speeds of each planet were converted into tones. Together, this attributed to enhancing knowledge of Kepler's Law of Planetary Motion.

Albert Einstein is a well-renown scientist who defined the theory of relativity. Yet, many may be unfamiliar with his connection to music. Einstein believed that 'life without playing music is inconceivable'. He started taking violin lessons at the age of five and fell in love with Mozart. Einstein said that 'Mozart's music is so pure and beautiful that I see it as a reflection of the inner beauty of the Universe'. Einstein used music to solve problems and worshipped Bach. He included the Bach Concerto for Two Violins in his book—*Einstein for the 21st Century: His Legacy in Science, Art and Modern Culture.*

When considering sound, it is a longitudinal wave defined as *a mechanical disturbance from a state of equilibrium that propagates through an elastic medium.* This means that unlike light, sound requires a solid, liquid or gas in order for it to travel. The physics behind how sound exists and its origination is quite interesting and rather extensive. Have you ever asked yourself 'where did sound come from?' The first sound to was the Big Bang—as the early Universe expanded, sound waves propagated through the dense medium that closed back on itself. As a result, the hypersphere of the Universe rang like a bell. The Big Bang Expansion has spanned over 13.77 billion years. Progressions from the Big Bang include changes in temperature, cosmic size, proton density, visibility distance, Universe size, sky brightness, etc. During that time, the Universe was small, hot, smooth and relatively dense—today, the Universe is considered to be large, cold, lumpy and relatively empty.

Acoustics is *the branch of physics that deals with the study of all mechanical waves in gases, liquids and solids.* How we describe sound and its array of types can vary based upon how it may be perceived, how it may make someone feel or a combination thereof. Sound may be described either objectively or subjectively. Scientists tend to be more 'quantitative' and use objective terminology, whereas musicians tend to be more 'sensory' and use more subjective terminology. Some subjective terminology includes pitch and loudness, while objective terms are frequency and waveform. There are different types of sounds including unpleasant, loud, soft, harsh, musical, and noisy. Both noise and music are combinations of sound waves with varying frequencies; however, noise is 'disordered' sound, whereas music is 'ordered' sound. Sounds may be audible, where it can easily be heard or inaudible where it cannot be heard.

9.2 Mechanics of music and cellular effects

Cells are able to communicate in numerous ways from electrical signaling to acoustic vibrations, and this can vary depending on the cell type. Because of how sound travels from the ear to the brain, neurons have commonly been studied when exploring acoustic properties. Some studies have suggested that music (~60 revolutions per minute, rpm) can initiate a process in the brain where it synchronizes with the beat, causing alpha brainwaves (~8–14 Hz). Other investigations have explored the musical impact on cell proliferation and viability (i.e. cancer cells and non-auditory cells), typically involving classical or jazz music. These reports suggest that music alters cell morphology and viability. There have been studies that have explored the musical impact that genres can have on cell cycle, cell growth and cell migration, as shown in figure 9.1. In some ways, these studies can be quite complex as different genres, time of exposure and frequencies can influence the cellular activities. An important aspect of sound is that genetically identical cells can behave differently, and variations in sounds can promote (or reduce) resistance to antibiotics and other drugs that may be used for treatment of disease. Having their own melody or vibration, numerous cells tend to have a distinctive response to varying sounds. For example, the heart has its own defined rhythm, blood has a distinct tempo, and the brain has its own perception of rhythm.

Cellular behavior is not distinctive from human emotion. It is well established that music can affect a person's emotional state, whether it is a positive or negative effect. Generally, music that has a slower tempo tends to be more calming, relaxes muscles, and can minimize stress and anxiety. A common stress biomarker is cortisol (also known as hydrocortisone), which is a glucocorticoid generated and secreted by the adrenal cortex. Typically increased in response to stress, cortisol contributes to glucose homeostasis, inflammation, and immunosuppression. Music has revealed that it can lower the expression of cortisol, ultimately reducing stress; it is also thought to repair neurons and facilitate neurogenesis.

The unique sounds that a person regularly hears are important. Whether it is a parent hearing the voice or cries of their child, a mechanic knowing the sounds of an engine, the 'cha-ching' sound of a cash register or the sound of the refrigerator in a person's house, these sounds are significant. With the changes related to the COVID-19 pandemic, even professional sports such as the National Basketball Association (NBA), Major League Baseball (MLB) and National Hockey League (NHL) have felt pressure to establish 'natural' sounds for each of their respective sporting events. Called 'Sounds of the Game', sound engineers for the NBA have

Figure 9.1. Cultured cells treated with music.

developed and implemented sounds of fans that the players typically would hear during a game since spectators are not allowed. Even something as simple as a fan booing has been incorporated as a part of the sound noise. The fact is that sounds matter.

Besides impacting the cells and emotions, music has been used to improve knowledge and understanding. For instance, there are been some people that have developed songs to explain complex topics in science, technology, engineering, mathematics (STEM). Rhyme and rhythm can be great for enhancing memory, knowledge and understanding of more difficult topics, making it easier to digest and remember. While these studies are ongoing, there is data to support that 'content-based STEM music' lowers stress, increases student enjoyment for learning, and promotes long-term comprehension.

9.3 Frequency, waves and vibrations

The sine wave is the fundamental building block of all sounds, and has a unique set of anatomy. As shown in figure 9.2, the top of the wave or the peak is the **crest**, while the bottom of the wave is the **trough**. The distance from one crest (or trough) of a wave to another is defined as a **wavelength** (λ, lambda). The maximum displacement is the **amplitude**. In chapter 2, mechanics were introduced and more specifically kinematics. Because sound is a wave, it has the capacity to migrate. Thus, this movement can be quantified. Knowing that speed is the ratio of distance and time, this can translate to wave speed (v). Wave speed is a measure of the wavelength per time; the time (in seconds) required to complete one cycle is a **period** (T). Considering, wave speed is the ratio of wavelength and period. This can be demonstrated as

$$v = \frac{\lambda}{T}$$

Frequency (f) is the number of cycles completed per second. Measured in hertz (Hz), frequency is the inverse of a **period**.

$$f = \frac{1}{T}$$

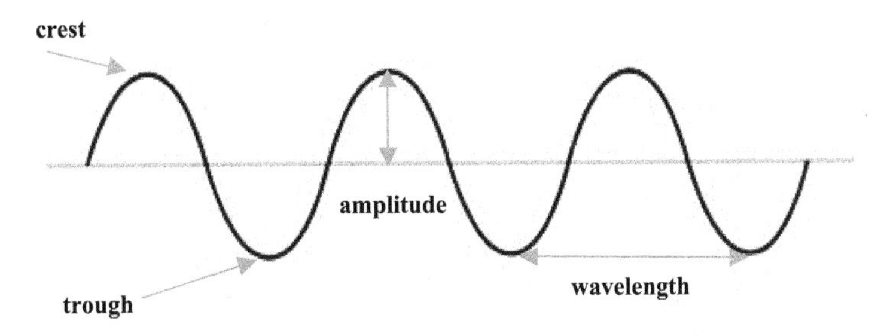

Figure 9.2. Schematic representation of a sine wave.

$$T = \frac{1}{f}$$

When translating frequency to wave speed,

$$v = \lambda f$$

Loudness relates to the intensity (or the amount of power output per area) of a sound wave, whereas pitch correlates with frequency. As a human, the normal range of hearing is 20 Hz to 20 000 Hz; yet, this range can vary depending on the species as well as with age. Frequencies that are below 20 Hz are known as **infrasonic**, while values above 20 000 Hz are **ultrasonic**. Table 9.1 shows a list of frequencies based on an assortment of musical instruments. As shown, all instruments have a distinctive frequency, where there is a fundamental frequency (or ground note) and a series of harmonics and overtones. In general, instruments that have a low frequency tend to have a low pitch (and vice versa). To produce their musical sounds, most musical instruments depend on standing waves, which occur when both ends of a string are fixed. Natural frequencies are the frequencies where standing waves are produced. The lowest frequency of vibration of a standing wave represents a **fundamental frequency** (f_1)—this also accounts for the first harmonic of a standing wave. The human whole-body fundamental frequency ranges between 5 and 10 Hz. Series of frequencies that are multiples of the fundamental frequency or simply 'repeating signals' are **harmonics**. An **overtone** is a higher tone above the fundamental frequency. Thus, a second harmonic represents a first overtone, while a third harmonic is a

Table 9.1. Assortment of instruments and their corresponding frequencies.

Instrument	Frequency
Acoustic Guitar	200–300 Hz
Bass Guitar	16–256 Hz
Cello	65–988 Hz
Clarinet	125–2000 Hz
Drum	250–3500 Hz
Flute	250–2500 Hz
Piano	28–4100 Hz
Saxophone	55–1000 Hz
Timpani	75–200 Hz
Trombone	60–630 Hz
Trumpet	170–1000 Hz
Tuba	45–375 Hz
Violin	200–400 Hz
Voice	80–1000 Hz
Xylophone	700–3500 Hz

second overtone, and so forth. This is shown in example 9.1, where these frequencies can be quantified based on the following:

$$f_n = \frac{v}{\lambda_n} = n\frac{v}{2L} = nf_1$$

$n = 1, 2, 3, \ldots$ (n refers to the number of the harmonic).

Example 9.1. While singing his favorite song, James has a guitarist playing in the background (figure 9.3). If the length of the guitar strings is 0.81 m and the frequency is 256 Hz, what is the wave speed at the 3rd harmonic?

Solution: Let's consider what we are given:

$$L = 0.81 \text{ m}$$

$$f_3 = 256 \text{ Hz}$$

$$n = 3$$

$$v = ?$$

Considering, use

$$f_n = n\frac{v}{2L}$$

Plugging in the known values to solve for 'v',

$$256 \text{ Hz} = (3)\frac{v}{2(0.81 \text{ m})}$$

$$\mathbf{138.24\ m\ s^{-1}} = v$$

A vibration is denoted as '*a wiggle in time*'. Thus, the normal vibration of the body is critical when maintaining overall functional health. The human body's vibrational frequency ranges from 3 to 17 Hz, having a 'sensitive' range between

Figure 9.3. Singer with guitarist playing in the background

6 and 8 Hz. This range accounts for vital parts of the body, and can be modified by a person's thoughts, food and water intake, as well as physical activity. Some frequencies can influence brainwaves by encouraging healing of the body and mind. Whole body vibration (WBV) is a technique where the machine transfers energy to the body, ultimately forcing a contraction–relaxation sequence each second on the muscles. This technique dates back to the ancient Greeks who thought that 'shaking' the body would instigate a more rapid healing process. During the 1960s, Russian scientists were excited about vibration therapy, as it was similar to rhythmic neuromuscular stimulation. It has the capacity to vibrate the entire body, ultimately improving blood flow and aiding to regulate and normalize the body's synergy. WBV vibrates at a particular frequency, where the most effective frequency for many people is believed to be 60 Hz. It also stimulates bone regeneration, improves circulation, promotes weight loss, flexibility, enhances dopamine levels to improve brain function, and reduce cortisol levels. An additional technique is ultrasonic cavitation. It is distinctive in the fact that it is a body contouring method that uses ultrasonic radio waves to disrupt fat cells. This is achieved by ultrasound waves traveling deep into the layers of the skin, dislodging fat cells from the dermal layers. This process generally utilizes devices that disburse 4–8 megahertz (MHz) of ultrasound, over a time frame of 20–60 min.

9.4 Beats, resonance and the Doppler effect

'The pulsations of sound intensity caused by the periodic coincidence of the amplitudes of two sound waves of slightly different frequencies' are called **beats**. It is believed that beats are the musical arrangement of a song or simply the instrumentals of a song without the lyrics. It is very common for someone to listen to music or a particular song based upon the beats within the song. Lower beats with lyrics can be more attractive as it draws you to the lyrics more. Alternatively, being familiar with particular musical artists can be more striking and empowering. Have you ever listened to your favorite song or artist and it uplifts you? Furthermore, has it inspired you to sing the words to the song (even if you may not be able to carry a tune)? This is the power of beats. **Beat frequency** is the difference between two frequencies and is best denoted as:

$$f_B = f_2 - f_1$$

For example, if two frequencies are provided, there will only be one beat frequency; however, if there are three or more beat frequencies, the number of possible beat frequencies will vary based upon the number of frequencies provided, as shown in example 9.2.

Example 9.2. Noah was listening to a song from his favorite hip-hop artist. Throughout the song, there were a series of beats that had frequencies of 440 Hz, 432 Hz, 512 Hz, and 528 Hz. Determine the beat frequencies possible.

Solution: Based on the four frequencies provided, the equation for beat frequency can be used to determine all of the possibilities:

$$f_B = f_2 - f_1$$

$$f_B = 440\text{ Hz} - 432\text{ Hz} = \mathbf{8\ Hz}$$

$$f_B = 528\text{ Hz} - 512\text{ Hz} = \mathbf{16\ Hz}$$

$$f_B = 512\text{ Hz} - 440\text{ Hz} = \mathbf{72\ Hz}$$

$$f_B = 512\text{ Hz} - 432\text{ Hz} = \mathbf{80\ Hz}$$

$$f_B = 528\text{ Hz} - 440\text{ Hz} = \mathbf{88\ Hz}$$

$$f_B = 528\text{ Hz} - 432\text{ Hz} = \mathbf{96\ Hz}$$

The superposition principle is when two or more waves occupy the same space at the same time, resulting in the algebraic sum of each wave's individual disturbance. This concept of interference can be separated into two aspects: **constructive interference**—when the peak (or crest) of one wave overlaps with the peak of another, leading to an increased amplitude; **destructive interference**—when the peak of one wave crosses over with the trough of another, resulting in a decrease in amplitude.

If two sources of sound are a certain distance from each other and one is sounded, the other sound will absorb energy and start vibrating at a frequency that matched the initial sound. This concept is known as **resonance**. Common examples of resonance include riding in a swing and even tuning in to your favorite radio station. *A natural frequency of vibration determined by the physical parameters of the vibrating object* is defined as a resonant frequency. In the aspect of the human body, resonance exists and can vary based on the orientation of the body as well as the individual. The heart resonance frequency is approximately 1 Hz.

Doppler effect is when *an increase (or decrease) in the frequency of sound occurs as a source and observer move toward (or away from) each other'*. This results in a sudden change in pitch and the frequency can be quantified by

$$f_o = \frac{v + v_o}{v + v_s} f_s$$

where f_o represents the frequency of the observer, v is the speed of sound, v_o is the observer's velocity, v_s is the velocity of the source, and f_s is the frequency of the sound waves. A perfect example would be if an ambulance were approaching you, the siren would be high pitched; however, it would lower as the ambulance traveled further away from you. As described in chapter 6, normal and abnormal blood flow through the heart can be measured with the echo that utilizes pulsed or continuous wave Doppler ultrasound.

9.5 Musical impact on disease

An anonymous author once said 'Music is a force that can unite humans even as they are separated by distance and culture'. In chapter 2, the concept of force was discussed simply being a push or a pull. Pleasant sounds are considered to be 'musical', and the body tends to naturally pull towards them. Yet, the variations in music can modify these effects. Music comes in a plethora of genres such as gospel, rock, country, classical, jazz, rhythm and blues (R&B), rap, pop, Afrobeats, funk, and the list can go on. In many instances, these genres can be categorized based on culture and typically elicit an emotional response. Depending on the individual, some musical genres can promote homeostasis in the body. In chapter 5, free radicals are introduced. Knowing that free radicals are 'robbers of energy', it should be no secret that they impact a number of diseases. Because of the innumerable side effects associated with medications, scientists often explore alternative treatments for varying diseases—music being one of them. Music therapy is *the utilization of music to address the cognitive, emotional, physical, and/or social needs of an* individual. Patterns and rhythmic contexts of music can improve cognitive functions. Sound healing is also a popular practice that uses audio tunes and vibrational frequencies to treat tissue and cellular damage in the body. Common diseases that benefit from music include cancer, multiple sclerosis (MS), diabetes, Parkinson's Disease, Alzheimer's Disease, and Dementia. Music therapy improves emotion and is very common in treatment for these diseases. More recently, correlations with COVID-19 and music have also been explored. Studies suggest that individuals with MS that were subjected to music experienced better word memory and word order in comparison to individuals that did not have music exposure. When it comes to type 2 diabetes, music can influence the main contributors to dysfunction: insulin resistance and hyperglycemia. These effects tend to have downstream detriments. Listening to music relaxes blood vessels and strengthens the immune system. Music also increases immunoglobulin A, which is an antibody that aids in attaching germs and bacteria that invade the body. There is evidence that music can lower blood glucose levels via the stress factor—levels of cortisol are reduced, while oxytocin is enhanced. Numerous studies suggest that music improves muscles and nerves. For example, individuals with Parkinson's disease tend to walk better in the presence of music. In general, the elderly tend to have better gait and balance when listening to music. Music has been used as a mnemonic to study gesture sequences in normal aging and Alzheimer's disease. Overall, Pythagoras has the right idea—music can be a natural, holistic treatment for diseased states and just the body as a whole.

Further reading

Brownjohn J M and Zheng X 2001 Discussion of human resonant frequency *Second Int. Conf. on Experimental Mechanics* vol 4317 (Bellingham, WA: SPIE) 469–74

Exbrayat J M and Brun C 2019 Some effects of sound and music on organisms and cells: a review *Ann. Res. Rev. Biol.* **32** 1–12

Feng R, Zhao Y, Zhu C and Mason T J 2002 Enhancement of ultrasonic cavitation yield by multi-frequency sonication *Ultrason. Sonochem.* **9** 231–36

Fukui H and Toyoshima K 2008 Music facilitate the neurogenesis, regeneration and repair of neurons *Med. Hypotheses* **71** 765–69

Geist K, Geist E A and Kuznik K 2012 The patterns of music *Young Child.* **2** 75

Kepler J 1619 *Harmonices Mundi Libri V* Book II, Sections 18–20 (Linz) pp 51–6

Koelsch S, Fuermetz J, Sack U, Bauer K, Hohenadel M, Wiegel M, Kaisers U and Heinke W 2011 Effects of music listening on cortisol levels and propofol consumption during spinal anesthesia *Front. Psychol.* **2** 58

Kučikienė D and Praninskienė R 2018 The impact of music on the bioelectrical oscillations of the brain *Acta Med. Litu.* **25** 101

Ladbury J E and Arold S T 2012 Noise in cellular signaling pathways: causes and effects *Trends Biochem. Sci.* **37** 173–8

Lestard N R and Capella M A 2016 Exposure to music alters cell viability and cell motility of human nonauditory cells in culture *Evid.-Based Complementary Altern. Med.* **2016** 6849473

Mehrafsar A and Mokhtari M J 2018 Effect of exposure to Quran recitation on cell viability, cell migration, and BCL2L12 gene expression of human prostate adenocarcinoma cell line in culture *Health, Spiritual. Med. Ethics* **5** 46–52

Moussard A, Bigand E, Belleville S and Peretz I 2014 Music as a mnemonic to learn gesture sequences in normal aging and Alzheimer's disease *Front. Human Neurosci.* **8** 294

Robb S L 2000 Music assisted progressive muscle relaxation, progressive muscle relaxation, music listening, and silence: a comparison of relaxation techniques *J. Music Therapy* **37** 2–21

White H E and White D H 2014 *Physics and Music: The Science of Musical Sound* (Chelmsford, MA: Courier Corporation)

Chapter 10

Power tools and the body

As defined in chapter 3, **power** is 'the amount of energy transformation over a period of time' and it aids in generating order in the body. When thinking of a **tool**, it is *an object used to extend the ability of an individual to modify features of the surrounding environment.* In general, a tool can sustain 'various degrees of force'; yet, tools can come in innumerable forms and have a variety of functions. A **power tool** is 'a tool that is triggered by an additional power source and mechanism other than the manual labor used with hand tools. Ultimately, these tools are primarily used externally and classified as stationary or portable. However, having a clear understanding of what 'power' and 'a tool' are allows a different perspective to be viewed of not only the external forces driven by the body, but the internal forces as well. One may ask the following: (i) what happens to the body when utilizing a standard power tool; (ii) what amount of energy is required to exert the forces needed to operate a power tool; (iii) what tools within the body drive the power process? These questions are explored throughout this chapter as we dive into the idea of essential external and internal power tools, as well as how they affect the various senses of the body.

10.1 The power tools that are useful for everyday life

When trying to achieve a simple task such as a home improvement or do-it-yourself (DIY) project, or something more complex, identifying the correct and most efficient power tool for a specific task is critical. A **power tool** is a tool operated by electricity or an electric motor. In the 1800s, power tools were invented; before this time, people did not have the necessary tools to complete tasks such as building a house or cutting down a tree. Power tools are designed and developed with a particular function in mind. Chapter 2 introduces the concept of torque. Torque is important in how a power tool functions and is utilized.

Typically, all power tools make a noticeable noise and you can usually feel some type of vibration/sensation. Electrical energy plays a vital role in the inner workings

doi:10.1088/978-0-7503-3283-5ch10

of power tools, helping to explain 'how' and 'why' various tools operate in the way that they do. Conventional flow is typical for how electricity follows a path through a power tool. The operations of power tools are explained by the 'power-path logic'. Power tends to flow from the power cord to the power switch to the brushes, armature and then the field. There are two main categories that power tools fall into: (a) electric or stationary; (b) battery-operated or portable. In general, the stationary power tools have more voltage in comparison to the portable tools. However, there are some power tools such as the drill and side grinder, that can be both electric and battery-operated. A side grinder is used to cut, grind, and polish both metals and nonmetals. Categories of power tools are highlighted in table 10.1.

Power tools can come in many different forms. The most common power tool is the drill, which has a variety of functions from fixing a loose screw to making a hole in a wall to hang something such as a painting. It can be used to create holes, remove broken components in metals or different objects. Either way, a drill is an essential power tool for many. It has many capabilities and various drill bits can be placed at the tip of the drill chuck, as shown in figure 10.1. Similar to a drill is an impact. The difference between a drill and an impact is that the impact has a 'snub' nose (see figure 10.2); it tends hold a bit better and provides more versatility such as being able

Table 10.1. Types of power tools and characteristics.

Type of power tool	Characteristics
Electric/stationary	Typically ~115 V Advantages: speed and precision Most common: miter saws, table saw
Battery operated/portable	Voltage can range from 6 to 48 V Hand-held and do not require an electric cord Most common: drills, impacts, chainsaw

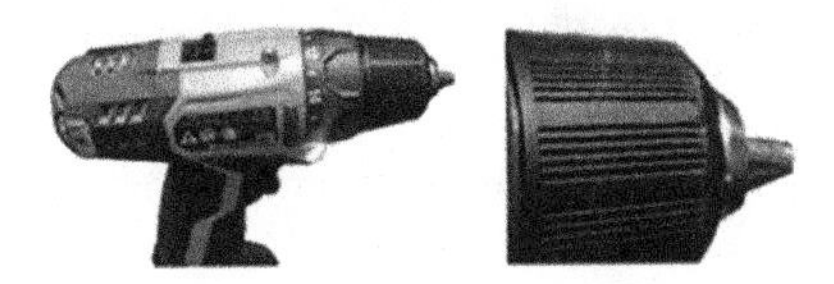

Figure 10.1. Drill. On the right is a zoomed in version of the drill chuck, highlighting the 'pointed' tip.

Figure 10.2. Impact. On the right is a zoomed in version of the impact chuck, highlighting the round tip.

to hold a 'Phillips head' bit. In general, an impact tends to have more torque and will run faster than the drill—it makes more of an 'impact'. Most impacts and drills use a lithium ion battery (those that are battery-operated).

The drill can also be used similar to the impact; however, the impact is more effective because the bit is designed for the impact. The drill in figure 10.1 is 12 V and 43 W·h and 4.0 A·h. The black portion on both the impact and the drill is called a chuck—this is where you insert your bits for the impact or the drill bits. Because of the tip, the impact will not hold a drill bit; however, a drill will hold bits for both a drill and an impact. Impact driver is similar to a drill but it is stronger, lighter, smaller, and best of all—safer. The driver creates both downward and rotational force, which assists with loosening screws and nuts.

An air compressor is necessary to power up tools such as jackhammers and nail guns; they are also commonly found in heaters, air conditioners and ventilators. An air compressor has energy that is compressed in the form of electrical energy or gasoline. Upon usage, the 'compressed energy' is stored until it reaches its brink. Heat guns require electricity to generate heat, where the air temperature can range anywhere from 100 °C to 550 °C. These tools are useful to shrink/melt materials, thaw frozen pipes and strip old paint.

There are also a plethora of saws that are power tools. A bandsaw is used to easily create equal lengths of a material with accuracy and precision (see figure 10.3), whereas a circular saw (see figure 10.4) tends to have sharper and more abrasive blades. The idea of the circular saw is to cut material in a rotary motion; it utilizes centripetal force, which was discussed in chapter 2. They are best known for their ability to accurately and efficiently perform straight cuts better than any other

Figure 10.3. Bandsaw. It provides great accuracy and precision.

Figure 10.4. Skill (circular) saw. It uses rotational force, and has sharp blades.

power tool. A miter saw is pulled straight down and is best for cutting at a particular angle due to it having a blade mounted to its swing arm that allows pivoting at varying angles. There are an array of miter saws such as compound miter saws, sliding compound miter saws, dual compound miter saws, etc. The variety provides different levels of 'tilting' based on the necessity of the task at hand. Scroll saws are ideal for delicate materials, making them desired by artists, crafters, quilters and tailors. As a machine-powered saw, the reciprocating saw is a general term for saws that have the ability to cut back and forth. These can include jigsaws, scroll saws, sabre saws, and rotary reciprocating saws. Reciprocating saws are commonly used for construction and demolition and achieve cutting via a 'push-and-pull' motion of the blade. These saws are differentiated from a jigsaw based on their cutting capabilities. Reciprocating saws mainly offer both horizontal and vertical cutting, whereas a jigsaw is more versatile and precise by also having the capacity to make bevel and compound cuts.

10.2 Internal tools that 'power' the body

Revisiting the definition of power and tool, it is evident that the dominant 'power tools' of the body are ones that perform work over time and modify the activities within the body. So what internal components contribute to such behavior? While the list can vary depending on one's perspective, these aspects are generally viewed mainly from a biological perspective; however, there is also biophysical importance associated with each of these tools as well. A few internal 'tools' are addressed and correlated to how and why they are considered as power tools. Together, they work together to regulate and power the body. Some of these tools include the brain, heart, liver, muscles, etc, all of which are described below.

Brain—Key aspects of the brain were addressed in chapter 7; however, the brain is the motor and operator of the body—it regulates networking. A variety of electrical processes are regulated by the brain, ultimately being one of the dominant internal power tools of the body. Without a functioning brain, there cannot be a functioning body.

Heart—As highlighted in chapter 6, the heart is critical in the body's circulatory system. Thus, it is no secret that this organ also plays an important role as a power tool within the body. The heart is considered as the battery of the body; the body is unable to perform essential tasks properly, without a fully functioning heart.

Liver—The liver is the largest internal organ in the body, playing a critical role in digestion, making proteins, and promoting toxicity regulation. Uniquely considered as both a gland and an organ, the liver breaks down substances in the body and functions as a storage unit. It is responsible for chemical actions that are important for survival. There is even a supplement on the market called 'Liver Power'. Liver Power is a combination of antioxidants, amino acids and herbs, having the ultimate goal to initiate detoxification.

Muscles—Muscles are important to both force and mechanics. In section 4 of chapter 2, bodily movements are addressed and the three types of muscles are highlighted: (i) cardiac; (ii) skeletal; and (iii) smooth. This aspect of the body takes a hit especially during weight training. Muscles are also critical when it comes to being able to properly lift, bend and move.

Blood—In chapter 6, the importance of the blood and blood flow is discussed as it relates to cardiac conductivity. The average adult produces approximately 2 L of blood per day. It powers the body through its ability to carry oxygen. Hemoglobin and hematocrit play a role, as well as iron. Think about it, your blood provides a lot of critical information about the health status of the body. Blood has also been uniquely used to fuel batteries via heme, which is a biomolecule from hemoglobin that carries oxygen in the blood. It is believed that these new batteries may allow electronic devices to be functional without needing to be charged for weeks at a time.

Stomach—This muscular organ is essential for acid secretion and enzymes for digestion. The pH of stomach acid is 1–3, making it slightly below the pH for a battery. There have been studies that demonstrated ways in which to utilize stomach acid as a source of power. Due to such a low pH, acid within the stomach is able to rapidly eat through food.

Eyes—Have you ever seen something and then desired it? The 'power' or ability to project electrical energy from a person's eyes seems rather supernatural; however, this concept is quite feasible. The eyes are considered as a power tool as they suggest what a person may be thinking, the next action they may take, or even provide insight as to a particular direction they may navigate. Previous work suggests that there is a force that emanates from the eyes. There is also a concept of what's called 'Substance C'—which is believed to model the events that are occurring in a person's head.

Taken together, both external or internal power tools have the ability to significantly impact the body and a person's life. Understanding the 'how' and 'why' for each of these aspects is vital in having a deeper understanding of the workings of the body.

Further reading

Hall J E and Hall M E 2020 *Guyton and Hall Textbook of Medical Physiology e-Book* (Amsterdam: Elsevier Health Sciences)

Mimee M *et al* 2018 An ingestible bacterial-electronic system to monitor gastrointestinal health *Science* **360** 915–8

Robson N and Wadge B 2003 Black and Decker Inc. Power tool *US Patent* 6,641,467

Tokunaga M and Ishikawa G 2005 Makita Corp. Power tools *US Patent* 6,968,908

https://homestratosphere.com/types-of-power-tools/

https://mycustomcleanse.com/liver-power.html

IOP Publishing

Biophysics of the Senses (Second Edition)

Tennille D Presley

Bibliography

Aprelev A W, Stephenson H, Noh M, Meier and Ferrone F A 2012 The physical foundation of vasoocclusion in sickle cell disease *Biophys. J.* **103** L38–40

Bagasra O, Michaels F H, Zheng Y M, Bobroski L E, Spitsin S V, Fu Z F, Tawadros R and Koprowski H 1995 Activation of the inducible form of nitric oxide synthase in brains of patients with multiple sclerosis *Proc. Natl Acad. Sci. USA* **92** 12041–5

Bahrami B, Lavie N and Rees G 2007 Attentional load modulates responses of human primary visual cortex to invisible stimuli *Curr. Biol.* **17** 509–13

Becker R O 1990 *Cross Currents: The Promise of Electromedicine* (Los Angeles, CA: Jeremy P. Tarcher)

Becker R O and Selden G 1985 *The Body Electric: Electromagnetism and the Foundation of Life* (New York: Quill)

Beckman J S and Koppenol W H 1996 Nitric oxide, superoxide and peroxynitrite: the good, the bad, and ugly *Am. J. Physiol.* **271** C1424–37

Belfield L T, Jeffers A B, Davis A T, Kavanagh K and Presley T D 2019 Biophysical regulation of thermodynamic effects on type 2 diabetes *AIP Conf. Proc.* **2109** 070003

Benov L 2001 How superoxide radical damages the cell *Protoplasma* **217** 33–6

Berry M J, Justus N W, Hauser J I, Case A H, Helms C C, Basu S, Rogers Z, Lewis M T and Miller G D 2015 Dietary nitrate supplementation improves exercise performance and decreases blood pressure in COPD patients *Nitric Oxide* **48** 22–30

Biewener A A 1992 *Biomechanics—Structures and Systems* (Oxford: Oxford University Press)

Brownjohn J M and Zheng X 2001 Discussion of human resonant frequency *Second Int. Conf. on Experimental Mechanics SPIE* **4317** 469–74

Cember H 1996 *Introduction to Health Physics* 3rd edn (New York: McGraw-Hill)

Davis L 2019 *Body Physics: Motion to Metabolism* (Corvallis, OR: Oregon State University)

Exbrayat J M and Brun C 2019 Some effects of sound and music on organisms and cells: a review *Annu. Res. Rev. Biol.* **32** 1–12

Feng R, Zhao Y, Zhu C and Mason T J 2002 Enhancement of ultrasonic cavitation yield by multi-frequency sonication *Ultrason. Sonochem.* **9** 231–6

Ferrier G R and Howlett S E 2001 Cardiac excitation-contraction coupling: role of membrane potential in regulation of contraction *Am. J. Physiol.-Heart Circ. Physiol.* **280** H1928–44

doi:10.1088/978-0-7503-3283-5ch11

Fish R and Geddes L 2003 *Medical and Bioengineering Aspects of Electrical Injuries* (Tucson, AZ: Lawyers & Judges Publishing Company, Inc.)

Fitts P M 1954 The information capacity of the human motor system in controlling the amplitude of movement *J. Exp. Psychol.* **47** 381

Fujiwara O and Ikawa T 2002 Numerical calculation of human-body capacitance by surface charge method *Electron. Commun.* **85** 1841–7

Fukui H and Toyoshima K 2008 Music facilitate the neurogenesis, regeneration and repair of neurons *Med. Hypotheses* **71** 765–9

Garrett A T, Goosens N G, Rehrer N J, Patterson M J, Harrison J, Sammut I and Cotter J D 2014 Short-term heat acclimation is effective and may be enhanced rather than impaired by dehydration *Am. J. Hum. Biol.* **26** 311–20

Geist K, Geist E A and Kuznik K 2012 The patterns of music *Young Child.* **2** 75

Giancoli D C 2016 *Physics: Principles with Applications* (Boston, MA: Pearson)

Guyton A C and Hall J E 2000 *Textbook of Medical Physiology* 10th edn (Philadelphia, PA: W.B. Saunders Company)

Hall J E and Hall M E 2020 *Guyton and Hall Textbook of Medical Physiology e-Book* (Amsterdam: Elsevier Health Sciences)

Herman I 2007 *Physics of the Human Body* (Berlin/Heidelberg: Springer)

Hewitt P 2015 *Conceptual Physics* 12th edn (Glenview, IL: Pearson)

Hunter S D, Dhindsa M S, Cunningham E, Tarumi T, Alkatan M, Nualnim N and Tanaka H 2016 Impact of hot yoga on arterial stiffness and quality of life in overweight/obese adults *J. Phys. Act. Health* **13** 1360–3

Johnson E A 1965 Touch display: a novel input/output device for computers *Electron. Lett.* **1** 219

Jones A M 2014 Dietary nitrate supplementation and exercise performance *Sports Med.* **44** S35–45

Kaufman W and Johnston F D 1943 The electrical conductivity of the tissues near the heart and its bearing on the distribution of the cardiac action currents *Am Heart J.* **26** 42

Kepler J 1619 *Harmonices Mundi Libri V* Book II, Sections 18–20 (Linz) pp 51–6

Koch C and Hepp K 2007 *The Relation Between Quantum Mechanics and Higher Brain Functions: Lessons from Quantum Computation and Neurobiology* (Pasadena, CA: California Institute of Technology)

Koelsch S, Fuermetz J, Sack U, Bauer K, Hohenadel M, Wiegel M, Kaisers U and Heinke W 2011 Effects of music listening on cortisol levels and propofol consumption during spinal anesthesia *Front. Psychol.* **2** 58

Kučikienė D and Praninskienė R 2018 The impact of music on the bioelectrical oscillations of the brain *Acta Med. Litu.* **25** 101

Ladbury J E and Arold S T 2012 Noise in cellular signaling pathways: causes and effects *Trends Biochem. Sci.* **37** 173–8

Lestard N R and Capella M A 2016 Exposure to music alters cell viability and cell motility of human nonauditory cells in culture *Evid.-Based Complementary Altern. Med.* **2016** 6849473

Libby P and Theroux P 2005 Pathophysiology of coronary artery disease *Circulation* **111** 3481–8

Lundberg J O, Weitzberg E and Gladwin M T 2008 The nitrate-nitrite-nitric oxide pathway in physiology and therapeutics *Nat. Rev. Drug Discov.* **7** 156–67

Lundberg J O, Weitzberg E, Lundberg J M and Alving K 1994 Intragastric nitric oxide production in humans *Gut* **35** 1543–6

McCall R P 2010 *Physics of the Human Body* (Baltimore, MD: Johns Hopkins University Press)

Mehrafsar A and Mokhtari M J 2018 Effect of exposure to Quran recitation on cell viability, cell migration, and BCL2L12 gene expression of human prostate adenocarcinoma cell line in culture *Health, Spiritual. Med. Ethics* **5** 46–52

Miller G D, Marsh A P, Dove R W, Beavers D, Presley T D, Helms C, Bechtold E, King S B and Kim-Shapiro D B 2012 Plasma nitrate and nitrite are increased by a high nitrate supplement, but not by high nitrate foods in older adults *Nutr. Res.* **32** 160–8

Mimee M *et al* 2018 An ingestible bacterial-electronic system to monitor gastrointestinal health *Science* **360** 915–8

Moussard A, Bigand E, Belleville S and Peretz I 2014 Music as a mnemonic to learn gesture sequences in normal aging and Alzheimer's disease *Front. Human Neurosci.* **8** 294

Munger M A *et al* 2020 Tetrodotoxin-sensitive neuronal-type Na+ channels: a novel and druggable target for prevention of atrial fibrillation *J. Am. Heart Assoc.* **9** e015119

Nelson P 2004 *Biological Physics* (New York: WH Freeman) pp 315–32

Pearson T, McArdle A and Jackson M J 2015 Nitric oxide availability is increased in contracting skeletal muscle from aged mice, but does not differentially decrease muscle superoxide *Free Radic. Biol. Med.* **78** 82–8

Pereira C, Ferreira N R, Rocha B S, Barbosa R M and Laranjinha J 2013 The redox interplay between nitrite and nitric oxide: from the gut to the brain *Redox Biol.* **1** 276–84

Presley T D *et al* 2011 Acute effect of a high nitrate diet on brain perfusion in older adults *Nitric Oxide* **24** 34–42

Presley T D, Vedam K, Druhan L J and Ilangovan G 2010 Hyperthermia-induced HSP90-ENOS preserves mitochondrial respiration in hyperglycemic endothelial cells by down-regulating glut-1 and up-regulating G6PD activity *J. Biol. Chem.* **285** 38194–203

Robb S L 2000 Music assisted progressive muscle relaxation, progressive muscle relaxation, music listening, and silence: a comparison of relaxation techniques *J. Music Ther.* **37** 2–21

Roberts T J and Azizi E 2011 Flexible mechanisms: the diverse roles of biological springs in vertebrate movement *J. Exp. Biol.* **214** 353–61

Robson N and Wadge B 2003 Black and Decker Inc. Power Tool *US Patent* 6,641,467

Rosen G M, Britifan B E, Halpern H J and Pou S 1999 *Free Radicals: Biology and Detection by Spin Trapping* (Oxford: Oxford University Press)

Rush S, Abildskov J A and McFee R 1963 Resistivity of body tissues at low frequencies *Circ. Res.* **12** 40–50

Salisbury D and Bronas U 2015 Reactive oxygen and nitrogen species: impact on endothelial dysfunction *Nurs. Res.* **64** 53–66

Schwan H P and Kay C F 1957 Capacitive properties of body tissues *Circ. Res.* **5** 439–43

Shiva S 2013 Nitrite: a physiological store of nitric oxide and modulator of mitochondrial function *Redox Biol.* **1** 40–4

Stinstra J G, Hopenfeld B and MacLeod R S 2005 On the passive cardiac conductivity *Ann. Biomed. Eng.* **33** 1743–51

Tartt A N, Fulmore C A, Liu Y, Rosoklija G B, Dwork A J, Arango V, Hen R, Mann J J and Boldrini M 2018 Considerations for assessing the extent of hippocampal neurogenesis in the adult and aging human brain *Cell Stem Cell* **23** 782–3

Tokunaga M and Ishikawa G 2005 Makita Corp. Power Tools *US Patent* 6,968,908

Tytell M and Hooper P L 2001 Heat shock proteins: new keys to the development of cytoprotective therapies *Emerg. Therap. Targets* **5** 267–87

Wang Z, Deurenberg P, Wang W, Pietrobelli A, Baumgartner R N and Heymsfield S B 1999 Hydration of fat-free body mass: review and critique of a classic body composition constant *Am. J. Clin. Nutr.* **69** 833–41

Weyer S, Ulbrich M and Leonhardt S 2013 A model-based approach for analysis of intracellular resistance variation due to body posture on bioimpedance measurements *J. Phys. Conf. Ser.* **434** 012003

White H E and White D H 2014 *Physics and Music: The Science of Musical Sound* (Chelmsford, MA: Courier Corporation)

www.caymanchemical.com

www.eicom-usa.com

https://www.homestratosphere.com/types-of-power-tools/

http://livescience.com/37807-brain-is-not-quantum-computer.html

https://mycustomcleanse.com/liver-power.html

www.physicsclassroom.com

www.seiko-cleanenergy.com

CPSIA information can be obtained
at www.ICGtesting.com
Printed in the USA
BVHW062003241021
619762BV00003B/88